THE WORLDWIDE LISTENING GUIDE

by:

John A. Figliozzi

On the Cover:

The Eiffel Tower is one of the world's most-recognized landmarks, and served as one the first radio antenna towers in the world. It was built as an attraction for the World's Fair of 1889, which was held to celebrate the French Revolution. Two engineers who worked for Gustave Eiffel's bridge-building firm— Emile Nouguier and Maurice Koechlin—originally conceived the idea for the Tower. Construction required 26 months, and it was roundly criticized while being built. When it was completed, however, the Eiffel Tower instantly became a popular success and more than 2 million people visited the Tower during the 1889 World's Fair. More than 175 million people have visited the famous landmark since it first opened on March 31, 1889.

The Eiffel Tower was only supposed to remain in place for 20 years. But Eiffel, searching for ways to make it more permanent, kept using it for various types of scientific research. Following the success of the first radio signals broadcast to the Pantheon by Eugàne Ducretet in 1898, Eiffel approached French military authorities in 1901 with a proposal to make the Tower into a long-distance radio antenna. In 1903, a radio connection was made with the military bases around Paris and then a year later with the East of France. A permanent radio station was installed in the Tower in 1906, ensuring its continuing survival. Eiffel lived long enough to hear the first European public radio broadcast from an aerial on the Tower in 1921. He died on December 27, 1923, at the age of 91.

Among his other achievements, Eiffel served the Chief Engineer for lock construction during France's failed attempt to build the Panama Canal, and he designed the unique structure for the Statue of Liberty.

The Eiffel Tower is 318.7 meters (1045 feet) tall, weighs 7,300 tons, is constructed of 18,038 pieces of puddle iron held together with 2.5 million rivets, and is painted every 7 years with 50 tons of paint.

To learn more, visit the official web site of the Eiffel Tower: www.tour-eiffel.fr.

This book was developed by:
John A. Figliozzi
© Copyright 2008 John A. Figliozzi
All Rights Reserved

Published for Mr. Figliozzi by:
Master Publishing, Inc.
6019 W. Howard Street
Niles, IL 60714
voice: (847) 763-0916
fax: (847) 763-0918
e-mail: masterpubl@aol.com

Acknowledgements:
The ongoing support and friendship of the following people have been instrumental to the success of this project and their contributions are acknowledged here: Harold Sellers of the Ontario DX Association (ODXA); Bob Brown, Rich Cuff, Kris Field, Fred Kohlbrenner and Ralph Brandi of the North American Shortwave Association (NASWA); Tom Sundstrom of TRS Consultants; Ian MacFarland of Marbian Productions, Inc., formerly of Radio Japan and Radio Canada International; Jonathan Marks of Radio Netherlands; Steve Shaye; and last, but certainly not least, my family — my wife, Patricia; daughter, Jennifer; sons, Andrew, Brett and Jason.

A similar listing was previously published as *The Shortwave Radioguide* by the following Shortwave Clubs:

North American Shortwave Association
45 Wildflower Road
Levittown, PA 19057 USA

Ontario DX Association
P.O. Box 161, Station A
Willowdale, ON M2N 5S8 Canada

Printing by:
Arby Graphic Service, Inc.
6019 W. Howard Street
Niles, IL 60714
voice: (847) 763-0900
fax: (847) 763-0918

REGARDING THESE BOOK MATERIALS

Fourth Edition
5   4   3   2   1

Table of Contents

Introduction

As I sit down to begin writing this new edition of my book, I am listening to WWL 870 New Orleans as Hurricane Gustav bears down on the Gulf Coast. Only, I'm in upstate New York, not in Louisiana. It's the middle of the day, not the middle of the night. And on my WiFi radio, reception is perfectly clear – unaffected by static, distant signal fading, or adjacent channel interference.

I love radio. I assume that if you're holding this book, you do too. We probably share a youthful first experience with a small and much-cherished AM transistor radio, along with an early discovery that at night the number of stations that we could hear increased markedly. It didn't take too much longer for either of us to learn of shortwave and, in so doing, further indulge our growing fascination with things distant, foreign, diverse – and wireless.

I mention "wireless" because that seems to be a hurdle for a lot of radio aficionados who are having difficulty as the Internet Age gathers momentum. After all, what is more characteristic of traditional radio reception than "the ether," that somewhat mysterious electromagnetic ocean upon which "the waves" have always traveled to us?

Until the advent of radio, international content mostly stopped at national borders for commercial, political, language or provincial attitudinal reasons. Radio punched through those barriers thanks to the shortwave platform. Shortwave radio alone could feed an appetite for a form of international mass communication between nations and cultures. For that reason, shortwave became synonymous with international radio and vice-versa. It stayed that way for a very long time.

However, if one looks even just at the history of shortwave, there has been a continuum of development in the medium with the incorporation of satellites and relay stations into the effort to be better and more reliably heard. At the time, there were "purists" who decried those efforts, too.

Even though shortwave radio held a long monopoly as the sole platform capable of delivering content internationally, it has always been a relatively minor and minority interest, except in areas where diversity in thought and creativity was not tolerated or otherwise available. In hindsight, the technical and aural shortcomings of shortwave did not prevent it from achieving a critical mass of utility so long as there was a Cold War and no other technology emerged to challenge the medium's supremacy.

But the Berlin Wall came down, budgets for the BBC, Voice of America, and other broadcasters were cut, and new content delivery platforms were – and continue to be – developed. Many of these new platforms revolve around personal computers and the Internet, as well as the development of subscription satellite radio. And so today, we have several platforms and methods capable of disseminating radio content internationally.

All program makers create content to be heard by as many ears as possible, whether generally or within a more targeted audience. Clearly, the "platform" is not the message. It's a means of delivering the message. In the end, how much difference is there between the final relay to my receiver being a modem and WiFi router instead of a shortwave transmitter in Sackville or Bonaire?

The parameters of what can be termed "radio" have spread so wide that this book is going to have to set its own parameters. There are an estimated 20,000 radio stations alone streaming over the internet. And then there's hundreds of channels on Sirius XM. And, of course, there's traditional analog AM, FM, and shortwave radio. So covering them all in a book is just not practical.

My goal in writing *The Worldwide Listening Guide* is to help you improve your listening experience and find programs of interest to you. One way to achieve that goal is to help you understand the new audio platforms that are available to help you expand your listening choices. The program listings in this new edition focus on English language international radio content akin to traditional shortwave radio that is primarily, but not exclusively, non-commercial in nature. The book concerns itself with your listening opportunities to and from and in North America over a range of delivery platforms including:

1. Analog shortwave broadcasts to North America.
2. Sirius XM satellite radio services that meet the aforementioned programming criteria.
3. Internet audio streams available via computer, WiFi radio and other means, that meet the criteria.
4. Some wide-area AM stations – formerly known as clear channel stations – that use high power and can be heard over wide regions of North America at night.
5. FM public radio stations that relay BBC World Service and World Radio Network programming.

In addition, this book has an indirect focus on podcasts and on-demand broadcasting inasmuch as the stations, services and networks covered in the book provide much of their content in those additional formats.

New delivery platforms create new opportunities for broadcaster and listener. Not long ago, it would have been impossible for me to hear a local station in New Orleans during the daylight hours at my home in New York. Even at night, it may have been impossible for me to hear WWL.

To be sure, not one of these delivery platforms is perfect by its self; nor does any one of them fully trump any of the others. However, used together in some personally developed, convenient and useful combination, they clearly enhance the listening experience and more readily serve the goal of the content creator – to reach as many ears as possible. They also more readily serve your objective as a listener, making it easier for you to assure that your ears are included in that audience.

Put simply, that is the philosophy and objective of this book. It's also what makes this book unique. I believe this new approach is what ties past editions of *The Worldwide Shortwave Listening Guide* that I've authored to this new iteration, and projects it into a more useful *Worldwide Listening Guide* for the present and future.

Best wishes and good listening!

John Figliozzi

Radio Listening Options – My, How They've Grown!

AM, FM and shortwave – for decades, these were the options we had when we wanted to listen to radio. Many of us were not even aware of shortwave!

Fast forward to the first decade of the 21st century and those options have been greatly expanded. AM, FM and shortwave remain prominently with us, but their singular and longstanding analog delivery structure is being layered over with new digital formats for all three. A further digital option – subscription satellite radio – has been adopted by more than 20 million customers and is growing. In excess of 20,000 radio stations worldwide are using the internet to serve listeners well beyond the small, local areas their transmitters can reach. Other new audio content providers are eschewing the traditional modes and methods to make their programs available directly to listeners via the internet. Traditional radio receivers are being augmented by computers, iPods, MP3 players, cellular mobile telephones and new internet-linked "wireless digital stereo content players" – or WiFi radios.

AM/FM/SW

Analog

The eldest technology has not been sitting still. Over the years, continuous improvements have been made to analog AM, FM and shortwave receivers yielding better quality and lower cost units that have become as close to ubiquitous around the world as any product ever has. Good quality, decently performing portable radios by mass manufacturers like Kaito, Sangean and Eton, that include all three bands can be purchased for as little as $50US or less.

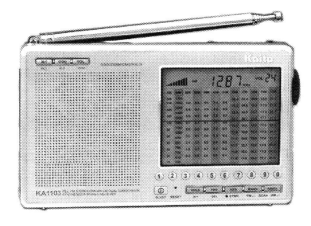

The Kaito KA1103 is a dual-conversion, portable AM, shortwave and FM radio. It has lots of useful features, is easy to use, and is relatively inexpensive.

As with all consumer products, however, performance improves and sophistication increases as one pays more. Some listeners prefer one band in particular and there are products that recognize and seek to serve that interest to the greatest extent that the technology permits. For example, one can pay in excess of $120US for a top of the line AM broadcast band radio, such as the CC Radio Plus, that specializes in long distance reception. The consensus best performing, most sophisticated (and, therefore, expensive) "all-in-one" AM-FM-SW portable receiver is the Eton E1 which retails for around $400US, but is quite intuitive to operate with many helpful and useful features not found on less expensive radios. As one might expect with technology so long on the market with just as long a history of innovation and improvement, there are more than ample and affordable options for everyone from the casual listener to the professional monitor.

Digital

There are significant advantages in "going digital". Analog signals are subject to a number of factors that degrade audio performance – fading, atmospheric static, man-made noise. Because digital signals are not so encumbered, they sound better and their level of quality is consistent. Digital signals also can be multiplexed, allowing multiple services to use the same frequency. However, one problem with digital is that when the signal level and quality degrades to a certain point, due either to distance from the transmitting source or the introduction of interference of some sort, audio is lost completely. At least with analog, the listener can continue to follow the program notwithstanding the fact that comfort and enjoyment will be hampered to some extent. Digital receivers – at least at present – also consume considerably more energy than more efficient analog radios. Therefore, analog radios are more portable and much cheaper and more convenient to operate – especially in areas not readily accessible to a mains power source.

Radio stations in Canada are experimenting with a DAB (Digital Audio Broadcasting) format developed in Europe called *Eureka 147*, which uses an entirely new frequency spectrum called the "N" band (1452-1492 MHz.). This means that new receivers are needed to access the digital signals. At this time, services are available only in a few major Canadian cities, so radios, while available, are not as easy to find as their analog cousins. It is early, but it's fair to say that progress in getting consumers to "buy in" has been slower than expected.

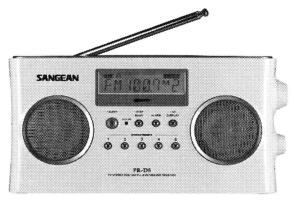

The Sangean WFR-20 WiFi Internet radio gives a listener access to over 6000 Internet Radio stations as well as 21,242 on-demand streams in 250 locations from 60 genres. And more!

The commercial radio industry in the United States opted to develop its own system, rather than adopt the Eureka 147 DAB standard to which most of the rest of the world has committed. *IBOC – in-band, on-channel* technology – was developed by a firm called Ibiquity and is being marketed as HD (or high definition) radio. A key feature and selling point for IBOC is the fact that the analog and digital signals share the same AM and FM frequency spectrum. This allows HD radios to revert back to analog reception on their primary channels if or when digital reception fades. Even so, as with Eureka 147, new radios (or converters) also are required in order to receive HD signals. The good news for consumers is that HD radios are showing up on more and more store shelves and they can be had for as little as $60US.

HD radio is being marketed to consumers by terrestrial radio stations as a subscription-free alternative to satellite radio. To a limited extent, HD radio does expand listening options. However, its development is being so cautiously managed by Ibiquity and the commercial radio industry in order to avoid adversely affecting the advertising revenues of existing stations, that innovation is being effectively choked. So far, the promise of HD radio is much less than it should be – certainly much less exciting than the era when FM first blossomed and reenergized radio in the late 1960s and early 1970s.

Digital Radio Mondiale (DRM) is shortwave radio's digital answer to the more serious aural shortcomings of its analog side. As exciting as hearing a station from halfway around the world on wireless can be for some listeners, for most others the pops, wheezes and static crashes that station picks up on the way from source to listener appears to serve as an impenetrable barrier for the enjoyment and embrace of the medium. Consequently, DRM seeks to put shortwave on a more equal technical quality footing with other, more accessible and acceptable (to the masses) media.

If any of these developing digital alternatives demonstrate the reality of a "digital divide" (a phrase describing the technological gap between richer and poorer societies) DRM probably serves as its poster child. The most vexing characteristics of digital technology today – its considerably higher energy consumption on the receiving end, its need for new and more expensive receivers, its discarding of older, ubiquitous and much more affordable radios, its attention to "values" more important to the wealthy (like pristine audio performance) than the poor (efficiency, utility and ease of access) – seem to undermine the most vital and baseline objectives of mass communication. Add to all this the fact that there is not yet on the consumer market a well-performing, affordable, "all-in-one-box" DRM receiver anywhere in the world and that only a handful of DRM broadcasts target North America on shortwave, and one can only come to the conclusion that DRM is still, at best, very much in an experimental phase of its development.

On DRM's behalf, when it does work it sounds marvelous, it uses the same frequency spectrum as analog AM, and it reduces costs for the broadcaster by trimming the transmission power required to deliver the signal. However, it clearly is not even a minor player yet.

SiriusXM Radio

North America's two subscription satellite radio services recently have merged. This is expected to yield new programming services, new product options and service configurations for consumers, along with improved, more flexible receivers. The fact that satellite radio has had the fastest and most successful roll-out of all consumer electronics products to date is directly attributable to the moribund state of affairs characterizing the programming available on much of terrestrial radio. Satellite radio has directly fed that listener yearning for more and varied options by programming for a greater variety of musical styles and tastes, wider topics of conversation, expanded news and international perspective and other interests ignored or underserved by commercial terrestrial radio.

The challenge for SiriusXM is not to make enthusiastic converts of radio listeners. In large measure, it's accomplished that already. What SiriusXM has to do next is demonstrate that the business model it ultimately settles on is at least viable and sustaining. As they say in the business, stay tuned.

The top-of-the-line Etón E1-XM is the world's first radio that combines AM, FM, shortwave and XM Satellite radio into one ultra high-performance unit.

Internet Audio

With as many as 20,000 radio stations and services worldwide available on the internet, broadcasters are demonstrating that they see both a present and a future for it as a delivery platform. Some even argue that it could supplant all the existing platforms. While that claim is clearly premature (and maybe far too grandiose), the internet is already at least an important adjunct and offers enhanced options for listening that heretofore had not existed. The internet makes computers, iPods, MP3 players and WiFi "digital stereo content players" into new and value-added forms of radio.

The internet also gives listeners some very flexible and user-friendly tools to time shift their listening. This is a win-win for both "content providers" (stations) and listeners. Stations use the internet, not only to "stream" their real time networks; but also to archive programming for listener on-demand access, either via another user-controlled stream or for download to a personal audio device like an iPod or MP3 player. These functions are starting to show up in cellular mobile phones, albeit at a price. It won't be long before manufacturers of other devices – some communications-based, others not – begin to see an audio application as a value added feature for their products, too.

That latter point is underlined by the fact that some functions that originated on the computer are beginning to migrate away from it and show up in other newly developed instruments on their own. One of these is the WiFi internet radio.

The Com One Phoenix Wi-Fi radio is definitely designed for a time when streaming Internet radio is listened to more than conventional AM/FM radio. The Phoenix works right out of the box, and the set up is very easy. The user has access to over 10,000 stations, as well as 5,000 podcasts that are provided by Podemus.

WiFi radio's most obvious benefit is that it addresses what has been the main complaint about computer-based internet radio – that it doesn't *feel* like radio. However, the WiFi radio has other compelling features, too. Yes, it can access the audio streams of thousands of radio stations. But it also can be set to automatically load and play podcasts, as well as serve as a remote player for music libraries stored on a PC hard drive. Many models also include an alarm clock feature that can help you drift off to sleep at night, wake you, and then allow you to snooze those precious extra few minutes in the morning.

Heretofore unknown names like *Revo, Tangent, Com One* and *Grace* have WiFi radios on the market based on the Linux operating system. So does a more familiar manufacturer, *Sangean*. A web-based (internet) interface is used to allow the consumer to customize each unit with a wealth of listening options and create his or her own organizational matrix. Most use technology developed by Reciva, which also provides an interface for the consumer as part of its licensing agreement with the manufacturer. Com One has developed its own proprietary interface for its Phoenix "digital stereo content player." In all cases, the consumer registers the unit and is duly rewarded with the ability to configure it to his or her listening needs. Otherwise, the interface is available but it's flexibility for the user is greatly truncated.

WiFi radios come preloaded with thousands of radio stations and podcasting streams. For desired content not already provided through the web interface, a facility is provided for testing audio stream and podcast links for compatibility with the player's firmware. Links to compatible content can then be added by the user into his or her personal account and then downloaded automatically through the internet to the player.

These preferences are then immediately and automatically communicated through the internet to the user's player. It's simpler and more intuitive in practice than it sounds from trying to explain it; but in the end it seems to address that primary criticism of internet-provided radio by bringing the "feel" of radio back to the foreground in the experience. The *Phoenix* and one *Revo* model – the *Pico* – operate on rechargeable batteries as well as on mains adapter.

The WiFi radio also addresses in a big way a fundamental yearning that radio content hounds like me have and that even shortwave can't satisfy entirely: that insatiable desire for more. There are stations available on the internet that have never been and will never be available in North America on shortwave. In short, it's an exciting new development. It should be interesting to see if it catches on with more listeners and consumers in North America as it apparently has in Europe.

The C Crane CC WiFi gives you a simple and economic way to enjoy Internet radio. The CC WiFi combines straightforward style with all the convenience of tuning in over 11,000 radio stations from around the world. The CC WiFi also has an Ethernet port for a hard-wire network connection.

Reviews and Advice

Evaluating the performance of various devices relative to one another goes beyond the immediate scope of this book. Fortunately, there are already some excellent sources for reviews and advice for shortwave and WiFi radios. Among them are:

- for shortwave radios, the annual guide **Passport to Worldband Radio** published every fall **www.passband.com** and **www.radiointel.com**.

- for WiFi radios, **www.wifiradioreview.com** and **www.cnet.com**.

So, Which Is Best?

In my opinion, the answer is all of them and none of them. Each has its advantages and drawbacks, depending on listening situations and what aspect of radio is emphasized.

- Does the broadcaster need unfettered direct access to the listener (and vice-versa)? Analog shortwave is still the best – and maybe only – solution.

- Do you want to listen to international programming content on your morning walk? Podcasting is the clearly superior mode for that. (Seriously, folks, carrying a portable shortwave radio with antenna fully extended while power walking can be dangerous, let alone cumbersome.)

- How about an insightful and entertaining conversation on RTE Ireland or a documentary from Radio Netherlands while driving to work? The World Radio Network via SiriusXM satellite radio can conveniently deliver that.

- Do you want to hear the BBC Proms concerts live in the best possible audio quality? The World Service or Radio 3 stream on the internet has to be the choice.

- Are you camping out in a remote area, need to keep the backpack light but still want to keep in touch with what's going on in the wider world. A cheap, small AM radio with decent skywave reception at night is all you need.

So, which is best? Every platform has its use. And that is why this book deals with all of them.

THE FOCUS OF THE WORLDWIDE LISTENING GUIDE

The parameters of what can be termed "radio" have spread so wide, that this book – if it's not going to resemble the metropolitan New York telephone directory – is going to have to set its own parameters. There are an estimated 20,000 radio stations alone streaming over the internet. Covering them all in a single book is just not practical. Here's what we're trying to accomplish with this book.

The Worldwide Listening Guide focuses on:

1. Listening to and from and in *North America*;

2. *Analog shortwave* to North America;

3. *English language international radio content* akin to traditional shortwave radio – *primarily but not exclusively non-commercial in nature* – with some additional programming that mostly meets or approximates that criteria;

4. *SiriusXM satellite radio* services that meet the aforementioned criteria;

5. *Internet audio streams*, available via computer, WiFI radio and other means, that meet the aforementioned criteria;

6. Some *wide area AM stations*, formerly known as clear channel stations, using high power and heard over wide regions of North America at night;

7. *FM public radio stations that relay BBC World Service and World Radio Network* programming.

In addition, this book has an indirect focus on podcasts and on-demand broadcasting inasmuch as the stations, services and networks covered in the book provide much of their content in those additional formats.

At this time, the WWLG does not focus any of its attention on:

A. DRM (Digital Radio Mondiale) broadcasts because (a) they are few in number, (b) targeted mostly to regions outside North America, (c) they are audible only with "homebrew" combination of computer, software, soundcard, receiver and antenna applications.

B. Most AM and FM broadcasts because they are primarily locally focused and too many in number to list, with much programming that replicates itself from locality to locality as part of one or more ad-hoc networks.

HOW TO LISTEN

Listening to the radio can mean a number of things today that it didn't mean just a few years ago. You can listen to radio…

… via an **analog AM/FM receiver**. (Yes, the one that receives relatively local stations and with which you're most familiar and probably use most frequently.)

… via an **HD and HD+-capable AM/FM receiver**. For HD and HD+ reception, you will need an entirely new radio – or at least a converter for your old one – that can tune into a station's digital signal and decode the additional digital channels it is broadcasting. In some cases, an enhanced antenna may be necessary to secure reception. A growing number of HD+ channels are carrying the *BBC World Service*.

… via a **shortwave radio receiver**. These radios have additional broadcast bands capable of receiving broadcast radio signals across intercontinental distances. Until quite recently, it was the only or most accessible means of doing so. Lately, international broadcasters have begun to de-emphasize or eliminate shortwave altogether as a delivery platform to North America.

… via a **satellite radio receiver**. A new form of radio introduced earlier this decade by Sirius and XM which have recently merged, it involves an entirely new and different set of receivers as well as a subscription agreement. It provides greater variety overall with additional and more focused channel formats. *BBC World Service, CBC Radio* and the *World Radio Network* services are available on SiriusXM satellite radio.

… via a **computer**. An ever increasing number of stations worldwide **"stream"** audio on the internet, effectively turning personal computers into radio receivers. Freely available software, usually already built-in, allows for "point and click" tuning that plays desired stations and content. The "listener" navigates to a station web site, or a web site that amalgamates links to a number of stations, and selects the desired **"audio stream"** or channel. Most stations and sites also offer convenience options that permit listening to particular programs at times and places of the user's own choosing through **on-demand** availability, **transfer of audio files** to individual hard drives, and **podcasting** – the automated download of content to hard drives for transfer to iPods, MP3 players and other mobile audio devices. The internet has increased the number of sources and options for listeners exponentially. This reality applies equally to **international broadcasting**.

… via a **Wi-Fi/Internet radio**. These devices are the latest innovation in receivers, allowing the listener to access audio content on the Internet in much the same way a common radio is used. WiFi radios gain access to a station's audio stream through a wireless internet router, bypassing both the Ethernet wire and the personal computer – although some models preserve the option to direct connect via an Ethernet wire. In addition, all the time-shifting convenience features detailed about computer listening are preserved in the Wi-Fi radio. A user can choose from a pre-selected roster of station links maintained by radio software providers like *Reciva, Tivoli* or *Com One*, depending on the device purchased. A personal catalog on a web site provided by the receiver manufacturer which connects to the radio registered on the site allows the user to further customize the content and operation of the receiver to preference. Full instructions are provided with each receiver.

The *Worldwide Listening Guide* recognizes this added flexibility and covers all of these listening options!

STATIONS LISTED IN THIS GUIDE

The following listing provides a key to the station symbols used in the Program List section of this Guide, as well as the name of the station, and the country where it is located.

Abbreviation	Name	Country
ABC-RA	Radio Australia	Australia
ABC-RN	ABC Radio National	Australia
BBC-R4	BBC Radio 4	United Kingdom
BBCWS-AM	BBC World Service Americas stream (XM Satellite Radio)	United Kingdom
BBCWS-IE	BBC World Service Info/Entertainment stream	United Kingdom
BBCWS-NX	BBC World Service News stream	United Kingdom
BBCWS-PR	BBC World Service U.S. Public Radio stream	United Kingdom
CBC-R1	CBC Radio 1 Atlantic Time Zone	Canada
CBC-R1C	CBC Radio 1 Central Time Zone	Canada
CBC-R1E	CBC Radio 1 Eastern Time Zone	Canada
CBC-R1M	CBC Radio 1 Mountain Time Zone	Canada
CBC-R1P	CBC Radio 1 Pacific Time Zone	Canada
CBC-R1S	CBC Radio 1 on Sirius Satellite Radio	Canada
CBC-RCI	CBC Radio Canada International	Canada
CRI-ENG	China Radio International English Service	China
CRI-RTC	China Radio International Round-the-Clock	China
CRI-WASH	China Radio International Washington	China
DW	Deutsche Welle	Germany
IRIB	Islamic Republic of Iran Broadcasting English Service	Iran
KBS-WR	Korea Broadcasting System World Radio	South Korea
NHK-RJ	Nippon Hoso Kyokai World Radio Japan	Japan
ORF-FM4	Osterreichischer Rundfunk fm4 network	Austria
PR-EXT	Polish Radio External Service	Poland
R.PRG	Radio Prague	Czech Republic
RAE	Radiodifusion Argentina al Exterior	Argentina
RAI-INT	Radiotelevisione Italiano (RAI) International	Italy
RFI	Radio France Internationale English Service	France
RNW	Radio Netherlands Worldwide	Netherlands
RNZ-NAT	Radio New Zealand National	New Zealand
RNZI	Radio New Zealand International	New Zealand
RTE-R1	RTE Radio 1	Ireland
RTE-R1X	RTE Radio 1 Extra	Ireland
RTHK-3	Radio Television Hong Kong Radio 3	China
RTI	Radio Taiwan International	China-Taiwan
SABC-CHAF	South African Broadcasting Corporation - Channel Africa	South Africa
SABC-SAFM	South African Broadcasting Corporation - SAfm	South Africa
VOR-WS	Voice of Russia World Service	Russia
WRN-NA	World Radio Network - North America	United Kingdom
WRS-SUI	World Radio Switzerland	Switzerland

Website Addresses of Stations/ Networks Included in the WWLG

Call Sign/Station ID	Name	Web Address
ABC-RAinet	Radio Australia English Language	www.radioaustralia.net.au/waystolisten/usa.htm
ABC-RNinet	ABC Radio National 2	www.abc.net.au/rn
BBC-R4inet	BBC Radio 4	www.bbc.co.uk/radio4/
BBCWS-IEinet	BBC W. S. Information & Entertainment	www.bbc.co.uk/worldservice/schedules/internet/wsradio_today.shtml
BBCWS-NXinet	BBC W. S. 24 Hour News Stream	www.bbc.co.uk/worldservice/schedules/internet/news_today.shtml
CBAMinet	CBAM Moncton	www.cbc.ca/listen/index.html
CBC-RCIinet	RCI Viva	www.rciviva.ca/rci/en/
CBCLinet	CBCL London	www.cbc.ca/listen/index.html
CBCSinet	CBCS Sudbury	www.cbc.ca/listen/index.html

CBCTinet	CBCT Charlottetown	www.cbc.ca/listen/index.html
CBCVinet	CBCV Victoria	www.cbc.ca/listen/index.html
CBDinet	CBD St. John	www.cbc.ca/listen/index.html
CBEinet	CBE Windsor	www.cbc.ca/listen/index.html
CBHAinet	CBHA Halifax	www.cbc.ca/listen/index.html
CBIinet	CBI Sydney	www.cbc.ca/listen/index.html
CBKinet	CBK Watrous	www.cbc.ca/listen/index.html
CBLAinet	CBLA Toronto	www.cbc.ca/listen/index.html
CBMEinet	CBME Montreal	www.cbc.ca/listen/index.html
CBNinet	CBN St. John's	www.cbc.ca/listen/index.html
CBOinet	CBO Ottawa	www.cbc.ca/listen/index.html
CBQTinet	CBQT Thunder Bay	www.cbc.ca/listen/index.html
CBRinet	CBR Calgary	www.cbc.ca/listen/index.html
CBTinet	CBT Grand Falls	www.cbc.ca/listen/index.html
CBTKinet	CBTK Kelowna	www.cbc.ca/listen/index.html
CBUinet	CBU Vancouver	www.cbc.ca/listen/index.html
CBVEinet	CBVE Quebec City	www.cbc.ca/listen/index.html
CBWinet	CBW Winnipeg	www.cbc.ca/listen/index.html
CBXinet	CBX Edmonton	www.cbc.ca/listen/index.html
CBYGinet	CBYG Prince George	www.cbc.ca/listen/index.html
CBYinet	CBY Corner Brook	www.cbc.ca/listen/index.html
CBZFinet	CBZF Fredericton	www.cbc.ca/listen/index.html
CFFBinet	CFFB Iqaluit	www.cbc.ca/listen/index.html
CFGBinet	CFBG Goose Bay	www.cbc.ca/listen/index.html
CFWHinet	CFWH Whitehorse	www.cbc.ca/listen/index.html
CFYKinet	CFYK Yellowknife	www.cbc.ca/listen/index.html
CHAKinet	CHAK Inuvik	www.cbc.ca/listen/index.html
CRI-RTCinet	CRI Round the Clock	http://english.cri.cn/
CRI-WASHinet	CRI Washington	http://english.cri.cn/webcast_washington/index.htm
DWinet	Deutsche Welle English Service	www.dw-world.de/
IRIBinet	IRIB studio 110 World Service	www.irib.ir/English/
KBS-WRinet	KBS World Radio	http://rki.kbs.co.kr
ORF-FM4inet	ORF-FM4 Live	http://fm4.orf.at/
PR-EXTinet	Radio Polonia - Polskie Radio SA	www.polskieradio.pl/zagranica/gb/
R.PRGinet	Radio Prague Live Broadcast	www.radio.cz/en/
RAEinet	Radiodifusion Argentina al Exterior	www.radionacional.com.ar/
RAI-INTinet	RAI International - Satel Radio	www.international.rai.it/radio/satelradio/index.shtml
RFIinet	RFI Multilingues 1	www.rfi.fr/actuen/pages/001/accueil.asp
RNW2inet	RNW2	www.radionetherlands.nl/#
RNZ-NATinet	Radio New Zealand National	www.radionz.co.nz/national/home
RNZinet	Radio New Zealand International	www.rnzi.com/pages/audio.php
RTE-R1inet	RTE Radio 1	www.rte.ie/radio1/index.html
RTE-R1Xinet	RTE Radio 1 Extra	www.rte.ie/radio1/index.html
RTHK-3inet	RTHK Radio 3 Live Broadcast	www.rthk.org.hk/channel/radio3/
CHAFStudio1inet	Channel Africa Live - Studio One	www.sabc.co.za/portal/site/channelafrica/
SABC-SAFMinet	SAfm	www.safm.co.za/portal/site/safm/
VOR-WSinet	Voice of Russia 2 Live	www.ruvr.ru/index.php?lng=eng
WRN-NAinet	WRN English for North America	www.wrn.org/listeners/stations/station.php?StationID=50
WRS-SUIinet	World Radio Switzerland	www.worldradio.ch

Looking to Explore WiFi Internet Radio Further?

Among the many and growing sources for internet radio, two stand out and are heartily recommended to you.

www.reciva.com – This site is maintained by Reciva, the company that developed and supports one of the primary chipsets used in WiFi internet radios. At last count, the number of sources it lists by country and genre was approaching 15,000.

www.publicradiofan.com – This is an amazing site created and maintained by Kevin A. Kelly that keys on public service radio around the world. It includes links and schedules organized in several intuitive ways. This is an entirely volunteer effort that greatly deserves the radio listening community's support in the form of encouragement, as well as information and cash contributions.

UTC Conversion Chart

The following chart shows the correct time conversion from UTC to each of the five times zones in the continental United States and Canada. You might want to copy this and keep it as a handy reference near your shortwave receiver and this Guide.

UTC	AST/EDT	EST/CDT	CST/MDT	MST/PDT	PST
0000	2000	1900	1800	1700	1600
0100	2100	2000	1900	1800	1700
0200	2200	2100	2000	1900	1800
0300	2300	2200	2100	2000	1900
0400	0000	2300	2200	2100	2000
0500	0100	0000	2300	2200	2100
0600	0200	0100	0000	2300	2200
0700	0300	0200	0100	0000	2300
0800	0400	0300	0200	0100	0000
0900	0500	0400	0300	0200	0100
1000	0600	0500	0400	0300	0200
1100	0700	0600	0500	0400	0300
1200	0800	0700	0600	0500	0400
1300	0900	0800	0700	0600	0500
1400	1000	0900	0800	0700	0600
1500	1100	1000	0900	0800	0700
1600	1200	1100	1000	0900	0800
1700	1300	1200	1100	1000	0900
1800	1400	1300	1200	1100	1000
1900	1500	1400	1300	1200	1100
2000	1600	1500	1400	1300	1200
2100	1700	1600	1500	1400	1300
2200	1800	1700	1600	1500	1400
2300	1900	1800	1700	1600	1500

AST — Atlantic Standard Time
EDT — Eastern Daylight Time
EST — Eastern Standard Time
CDT — Central Daylight Time
CST — Central Standard Time

MDT — Mountain Daylight Time
MST — Mountain Standard Time
PDT — Pacific Daylight Time
PST — Pacific Standard Time

How to Use the Program List

The Worldwide Listening Guide Program Lists contain more than 3,300 individual broadcasts that you can receive on your shortwave radio, SiriusXM radio, over the internet, or on your WiFi radio. To get the most from this list, you need to understand how it's organized and what all the numbers and symbols mean. The following is a guide explaining how to read the listings.

Time

The inclusion of program schedules that are organized by stations for international listeners in one case and domestic listeners in another complicate the task of determining the actual time a program will be available at the listener's location. Therefore, three time references are given for each program listing to assist you in pinpointing when it will be broadcast at different times of the year.

The entire list is organized by UTC (Universal Time Coordinated). The times begin at 0000 (midnight) and go through 2359 (11:59 p.m.). A table to convert UTC time to local time appears on page 12. In addition, as an aid in finding programs, the time range for programs listed on each page is shown at the upper outside corner of the page, similar to names in a phone book.

A second time reference – EDT or Eastern Daylight Time – is given in the next to last column of each program listing. Daylight time has been selected because it is now used eight months out of the year. Readers in other time zones have learned to adjust from EDT for their own locations.

A third time reference – Station Time – is given in the last column. With broader internet streaming of domestic stations, understanding when a program is intended to be heard at the station's home location may prove helpful.

Station / Network

The Station / Network IDs are the call signs or abbreviations for the various station. If you want to know the complete name of an individual station and its country of origin, check the Station / Network ID list on page 10.

Day(s)

The days of the week that a program is broadcast are represented by the numbers 1 – 7 (Sunday through Saturday). Slashes between the numbers indicate individual days, for example:

1/3/5 indicates that the program is broadcast on Sunday, Tuesday, and Thursday.

If a program is broadcast several days in a row, a dash between the numbers indicates the range, for example:

7-3 means that the program is broadcast Saturday through Tuesday.

Program Name

The program name column gives you the title of the program. A brief description of most programs is available by referring to the Program Descriptions section starting on page 84.

Type

The Type column contains the following codes indicating program content:

AC	Arts, Culture and History
BE	Business, Finance and Economic Development
CA	Current Affairs
CS	Society, Customs, Sites, Peoples and Cultural Values
DL	Everyday Domestic Live
DX	Media and Communications
ED	Ideas, Philosophy and Learning
EV	The Environment
GD	General Documentary Programs
GI	General Interest Programs
GL	Government, Politics and the Law
GZ	General Magazine-style Programs
HM	Health and Medicine
LD	Literature and Drama
LE	Light Entertainment, Humor, Quiz or Panel Game
LI	Listener Interaction or Mailbag Programs
LL	Language Lesson
M-	Music Programs
MC	Classical Music
MF	Folk and Traditional Music
MJ	Jazz and Blues Music
MP	Popular Music
MR	Rock Music
MV	Musical Variety
MX	General Music Program
MW	Country Music
MZ	World Music
MD	Music Documentary
N-	News Programming
NA	News Analysis
NC	News Commentary
ND	News Documentary
NX	General Newscast
NZ	News Magazine
PI	Personal Interview
PR	Press Review
SP	Sports Programming
ST	Science and Technology
TR	Travel and Tourism
VA	Variety
WX	Local Weather Report

Frequency / Platform

This column tells you how the program is delivered / available for your listening:

- Use of a **four or five digit number above 1710** indicates that the program is delivered via shortwave radio. Use of a **three or four digit number below 1711** indicates that the program is delivered via **AM radio**. All frequencies are given in kilohertz (kHz). So, if your receiver lists frequencies in megahertz (MHz) you must move the decimal point three placed to the left. 18100 kHz = 18.100 MHz.

- Use of the suffix **"inet"** after a series of letters indicates the program is delivered via **internet streaming**. Refer to the "Inet List" on page 10 for the internet address of the station and audio stream.

- Use of **"Sirius"** or **"XM"** followed by a number indicates that the program is delivered via **Sirius or XM satellite radio** on the channel number indicated.

- Use of **NPRfm/am** indicates that the program may be broadcast on **your local NPR affiliate's** AM or FM frequency. Use of **HD+** indicates that the program may be broadcast on one of your local **NPR affiliate's multiplexed HD channels**. Check you local station listings for confirmation.

Classified Program Listings

The Worldwide Listening Guide contains broadcasts that are targeted throughout the world and is organized primarily by time. To make it easier for you to find programs of interest to you, we have included the Classified Program Listings beginning on page 62. These lists are organized by 24 program types.

UTC Time Start	End	Station/ Network	Day(s)	Program Name	Type	Frequency/ Platform	EDT	Station Time
0000	0059	ABC-RA	1	The Spirit of Things	CS	RAinet	2000	1000
0000	0159	ABC-RA	2-6	Breakfast Club	VA	RAinet	2000	1000
0000	0029	ABC-RA	7	In the Loop	CA	RAinet	2000	1000
0000	0129	ABC-RN	1	Artworks	AC	ABC-RNinet	2000	1000
0000	0044	ABC-RN	2-6	The Book Show	AC	ABC-RNinet	2000	1000
0000	0159	ABC-RN	7	The Music Show	MC	ABC-RNinet	2000	1000
0000	0419	BBC-R4	1-7	(see BBCWS-IE)		BBC-R4inet	2000	0100
0000	0059	BBCWS-AM	1	The Strand	AC	XM131	2000	0100
0000	0059	BBCWS-AM	2	The Forum	CA	XM131	2000	0100
0000	0029	BBCWS-AM	3/5/7	Documentary feature or series	ND	XM131	2000	0100
0000	0029	BBCWS-AM	4	Global Business	BE	XM131	2000	0100
0000	0029	BBCWS-AM	6	Assignment	NA	XM131	2000	0100
0000	0019	BBCWS-IE	1/7	World Briefing	NX	BBCWS-IEinet	2000	0100
0000	0039	BBCWS-IE	2-6	World Briefing	NX	BBCWS-IEinet	2000	0100
0000	0019	BBCWS-NX	1/7	World Briefing	NX	BBCWS-NXinet	2000	0100
0000	0039	BBCWS-NX	2-6	World Briefing	NX	BBCWS-NXinet	2000	0100
0000	0019	BBCWS-PR	1/7	World Briefing	NX	Sirius141, NPRfm/am, HD+	2000	0100
0000	0039	BBCWS-PR	2-6	World Briefing	NX	Sirius141, NPRfm/am, HD+	2000	0100
0000	0059	CBC-R1A	2	Inside the Music	MZ	CBAMinet, CBTinet, CBDinet, CBHAinet, CBIinet, CBNinet, 640, 6160, CBTinet, CBYinet, CBZFinet, CFGBinet	2000	2100
0000	0059	CBC-R1A	3-7	Ideas	ED	CBAMinet, CBTinet, CBDinet, CBHAinet, CBIinet, CBNinet, 640, 6160, CBTinet, CBYinet, CBZFinet, CFGBinet	2000	2100
0000	0159	CBC-R1C	1	Vinyl Tap	MP	CBWinet, 990	2000	1900
0000	0059	CBC-R1E	2	Inside the Music	MZ	CBCLinet, CBCSinet, CBEinet, 1550, CBLAinet, CBMEinet, CBOinet, CBQTinet, CBVEinet, CFFBinet	2000	2000
0000	0044	CBC-R1E	4-7	The Night Time Review	CA	CBCLinet, CBCSinet, CBEinet, 1550, CBLAinet, CBMEinet, CBOinet, CBQTinet, CBVEinet, CFFBinet	2000	2000
0000	0029	CBC-R1M	1/2	The World This Weekend	NZ	CBKinet, 540, CBRinet, 1010, CBXinet, 740, CFYKinet, CHAKinet	2000	1800
0000	0029	CBC-R1M	3-7	The World At Six	NX	CBKinet, 540, CBRinet, 1010, CBXinet, 740, CFYKinet, CHAKinet	2000	1800
0000	0059	CBC-R1P	1	Regional performance	GI	CBCVinet, CBTKinet, CBUinet, 690, 6160, CBYGinet, CFWHinet	2000	1700
0000	0059	CBC-R1P	2	Writers and Company	AC	CBCVinet, CBTKinet, CBUinet, 690, 6160, CBYGinet, CFWHinet	2000	1700
0000	0059	CBC-R1S	2	Rewind	GI	Sirius137	2000	2000
0000	0059	CBC-R1S	3	And The Winner Is	FD	Sirius137	2000	2000
0000	0059	CBC-R1S	4	The Choice	GI	Sirius137	2000	2000
0000	0059	CBC-R1S	5	Dispatches	NA	Sirius137	2000	2000
0000	0059	CBC-R1S	6	Writers and Company	AC	Sirius137	2000	2000
0000	0059	CBC-R1S	7	The Vinyl Cafe	LE	Sirius137	2000	2000
0000	0019	CRI	1/7	News and Reports	NZ	6020, 9570	2000	0800
0000	0026	CRI	2-6	News and Reports	NZ	6020, 9570	2000	0800
0000	0029	CRI-RTC	1-7	News and Reports	NZ	CRI-RTCinet	2000	0800
0000	0005	DW	1-7	News	NX	DWinet	2000	0200
0000	0009	NHK-RJ	1-7	News	NX	6145	2000	0500
0000	0059	RNW	1	The State We're In	CS	RNW2inet	2000	0200
0000	0029	RNW	2/6	Network Europe Extra	AC	RNW2inet	2000	0200
0000	0029	RNW	3	Earthbeat	EV	RNW2inet	2000	0200
0000	0029	RNW	4	Bridges with Africa	CS	RNW2inet	2000	0200
0000	0029	RNW	5	Curious Orange	DL	RNW2inet	2000	0200
0000	0029	RNW	7	Radio Books	LD	RNW2inet	2000	0200
0000	0039	RNZ-NAT	1	Spectrum	DL	RNZ-NATinet	2000	1200
0000	0059	RNZ-NAT	2-6	Midday Report	NZ	RNZ-NATinet	2000	1200
0000	0159	RNZ-NAT	7	This Way Up with Simon Morton	EV	RNZ-NATinet	2000	1200
0000	1259	RNZI	1	(see RNZ-NAT)		RNZIinet, 15720, 9615s, 11725w, 7145s, 9655s	2000	1200
0000	0059	RNZI	2-6	(see RNZ-NAT)		RNZIinet, 15720	2000	1200
0000	0259	RNZI	7	(see RNZ-NAT)		RNZIinet, 15720	2000	1200
0000	0014	RTHK-3	1	RTHK News	NX	RTHK-3inet	2000	0800
0000	0029	RTHK-3	7	Today at Eight	NZ	RTHK-3inet	2000	0800
0000	0259	SABC-CHAF	1-7	African Music (News on the hour)	MF	CHAFStudio1inet	2000	0200
0000	0029	WRN-NA	1/2	Radio Australia features	VA	Sirius140, WRN-NAinet	2000	0100
0000	0014	WRN-NA	3-7	UN Radio: UN Today	NZ	Sirius140, WRN-NAinet	2000	0100
0005	0104	CBC-RCI	3-7	The Link (second hour)	GZ	6100, 9755, RCIinet	2005	2005
0005	0029	DW	1-7	Newslink	NZ	DWinet	2005	0205
0005	0009	RAI-INT	1-6	News Bulletin	NX	RAI-INTinet	2005	0205
0010	0029	NHK-RJ	1	World Interactive	LI	6145	2010	0510
0010	0029	NHK-RJ	2	Pop Up Japan	MP	6145	2010	0510
0010	0024	NHK-RJ	3/5/7	What's Up Japan	NZ	6145	2010	0510
0010	0029	NHK-RJ	4/6	What's Up Japan	NZ	6145	2010	0510
0015	0024	RTHK-3	1	Letter to Hong Kong	GL	RTHK-3inet	2015	0815
0015	0029	WRN-NA	3-7	Vatican Radio: News	NX	Sirius140, WRN-NAinet	2015	0115
0020	0029	BBCWS-IE	1/7	Sports Roundup	SP	BBCWS-IEinet	2020	0120
0020	0029	BBCWS-NX	1/7	Sports Roundup	SP	BBCWS-NXinet	2020	0120
0020	0029	BBCWS-PR	1	The Instant Guide	NA	Sirius141, NPRfm/am, HD+	2020	0120
0020	0029	BBCWS-PR	7	World Business Report	BE	Sirius141, NPRfm/am, HD+	2020	0120
0020	0026	CRI	1	Reports from Developing Countries	BE	6020, 9570	2020	0820
0020	0026	CRI	7	CRI Roundup	NX	6020, 9570	2020	0820

UTC Time Start	End	Station/ Network	Day(s)	Program Name	Type	Frequncy/ Platform	EDT	Station Time
0025	0029	NHK-RJ	3/5/7	Easy Japanese	LL	6145	2025	0525
0025	0029	RTHK-3	1	Thought for the Week	CS	RTHK-3inet	2025	0825
0027	0051	CRI-ENG	1	In the Spotlight	DL	6020, 9570	2027	0827
0027	0051	CRI-ENG	2	People in the Know	GL	6020, 9570	2027	0827
0027	0051	CRI-ENG	3	Biz China	BE	6020, 9570	2027	0827
0027	0051	CRI-ENG	4	China Horizons	DL	6020, 9570	2027	0827
0027	0051	CRI-ENG	5	Voices from Other Lands	DL	6020, 9570	2027	0827
0027	0051	CRI-ENG	6	Life in China	DL	6020, 9570	2027	0827
0027	0051	CRI-ENG	7	Listener's Garden	LI	6020, 9570	2027	0827
0030	0059	ABC-RA	7	Australian Bite	DL	RAinet	2030	1030
0030	0059	BBCWS-AM	3-7	The Strand	CS	XM131	2030	0130
0030	0059	BBCWS-IE	1	Politics UK	GL	BBCWS-IEinet	2030	0130
0030	0059	BBCWS-IE	7	Business Weekly	BE	BBCWS-IEinet	2000	0100
0030	0059	BBCWS-NX	1	The Interview	PI	BBCWS-NXinet	2030	0130
0030	0059	BBCWS-NX	7	Politics UK	GL	BBCWS-NXinet	2030	0130
0030	0059	BBCWS-PR	1	Reporting Religion	CS	Sirius141, NPR fm/am, HD+	2030	0130
0030	0059	BBCWS-PR	7	Heart and Soul	CS	Sirius141, NPR fm/am, HD+	2030	0130
0030	0059	CBC-R1C	2	C'est la vie	DL	CBWinet, 990	2030	1930
0030	0129	CBC-R1C	3	Dispatches	NA	CBWinet, 990	2030	1930
0030	0059	CBC-R1E	2	The Debaters	LE	CBCLinet, CBCSinet, CBEinet, 1550, CBLAinet, CBMEinet, CBOinet, CBQTinet, CBVEinet, CFFBinet	2030	2030
0030	0059	CBC-R1M	1	Laugh Out Loud	LE	CBKinet, 540, CBRinet, 1010, CBXinet, 740, CFYKinet, CHAKinet	2030	1830
0030	0129	CBC-R1M	2	Dispatches	NA	CBKinet, 540, CBRinet, 1010, CBXinet, 740, CFYKinet, CHAKinet	2030	1830
0030	0129	CBC-R1M	3	As It Happens	NZ	CBKinet, 540, CBRinet, 1010, CBXinet, 740, CFYKinet, CHAKinet	2030	1830
0030	0159	CBC-R1M	4-7	As It Happens	NZ	CBKinet, 540, CBRinet, 1010, CBXinet, 740, CFYKinet, CHAKinet	2030	1830
0030	0059	CBC-R1S	1	WireTap	LE	Sirius137	2030	2030
0030	0059	CRI-RTC	1	China Horizons	DL	CRI-RTCinet	2030	0830
0030	0059	CRI-RTC	2	Frontline	GL	CRI-RTCinet	2030	0830
0030	0059	CRI-RTC	3	Biz China	BE	CRI-RTCinet	2030	0830
0030	0059	CRI-RTC	4	In the Spotlight	DL	CRI-RTCinet	2030	0830
0030	0059	CRI-RTC	5	Voices from Other Lands	DL	CRI-RTCinet	2030	0830
0030	0059	CRI-RTC	6	Life in China	DL	CRI-RTCinet	2030	0830
0030	0059	CRI-RTC	7	Listeners' Garden	LI	CRI-RTCinet	2030	0830
0030	0044	DW	1/2	Sports Report	SP	DWinet	2030	0230
0030	0059	DW	3	World in Progress	BE	DWinet	2030	0230
0030	0059	DW	4	Spectrum	ST	DWinet	2030	0230
0030	0059	DW	5	Money Talks	BE	DWinet	2030	0230
0030	0059	DW	6	Living Planet	EV	DWinet	2030	0230
0030	0059	DW	7	Inside Europe	NZ	DWinet	2030	0230
0030	0129	IRIB	1-7	Koran Reading, News and features	VA	IRIBinet	2030	0400
0030	0059	RNW	2	Radio Books	LD	RNW2inet	2030	0230
0030	0059	RNW	3	Curious Orange	DL	RNW2inet	2030	0230
0030	0059	RNW	4	The State We're In Midweek Edition	CS	RNW2inet	2030	0230
0030	0059	RNW	5	Radio Books	LD	RNW2inet	2030	0230
0030	0059	RNW	6	Earthbeat	EV	RNW2inet	2030	0230
0030	0059	RNW	7	Bridges with Africa	CS	RNW2inet	2030	0230
0030	0059	RTHK-3	1	Give Me Five	LE	RTHK-3inet	2030	0830
0030	0029	RTHK-3	2-6	Backchat	DL	RTHK-3inet	2030	0830
0030	0059	RTHK-3	7	The Week on Three	GI	RTHK-3inet	2030	0830
0030	0059	WRN-NA	1	World Vision	BE	Sirius140, WRN-NAinet	2030	0130
0030	0045	WRN-NA	2	Radio Guangdong: Guangdong Today	DL	Sirius140, WRN-NAinet	2030	0130
0030	0059	WRN-NA	3-7	Radio Slovakia International	VA	Sirius140, WRN-NAinet	2030	0130
0040	0049	BBCWS-IE	2-6	Analysis	NA	BBCWS-IEinet	2040	0140
0040	0049	BBCWS-NX	2-6	Analysis	NA	BBCWS-NXinet	2040	0140
0040	0049	BBCWS-PR	2-6	Analysis	NA	Sirius141, NPRfm/am, HD+	2040	0140
0040	0359	RNZ-NAT	1	The Arts on Sunday with Lynn Freeman	AC	RNZ-NATinet	2040	1240
0045	0059	ABC-RN	2-6	First Person	LD	ABC-RNinet	2045	1045
0045	0059	CBC-R1E	4-7	Outfront	DL	CBCLinet, CBCSinet, CBEinet, 1550, CBLAinet, CBMEinet, CBOinet, CBQTinet, CBVEinet, CFFBinet	2045	2045
0045	0059	DW	1	Inspired Minds	GI	DWinet	2045	0245
0045	0059	DW	2	Radio D	LL	DWinet	2045	0245
0045	0059	WRN-NA	2	UN Radio: The UN and Africa	NZ	Sirius140, WRN-NAinet	2045	0145
0050	0059	BBCWS-IE	2-6	Sports Roundup	SP	BBCWS-IEinet	2050	0150
0050	0059	BBCWS-NX	2-6	Sports Roundup	SP	BBCWS-NXinet	2050	0150
0050	0059	BBCWS-PR	2-6	Business Brief	BE	Sirius141, NPRfm/am, HD+	2050	0150
0052	0057	CRI-ENG	1-7	Learning Chinese Now	LL	6020, 9570	2052	0852
0100	0129	ABC-RA	1	In the Loop	CA	RAinet	2100	1100
0100	0133	ABC-RA	7	Total Rugby	SP	RAinet	2100	1100
0100	0159	ABC-RN	2-6	Bush Telegraph	DL	ABC-RNinet	2100	1100
0100	0129	BBCWS-AM	1	The World Today	MC	XM131	2100	0200
0100	0119	BBCWS-AM	2-6	World Briefing	NX	XM131	2100	0200
0100	0129	BBCWS-AM	7	From Our Own Correspondent	NA	XM131	2100	0200
0100	0129	BBCWS-IE	1	Global Business	BE	BBCWS-IEinet	2100	0200
0100	0119	BBCWS-IE	2-6	World Briefing	NX	BBCWS-IEinet	2100	0200
0100	0129	BBCWS-IE	7	From Our Own Correspondent	NA	BBCWS-IEinet	2100	0200
0100	0129	BBCWS-NX	1	Assignment	ND	BBCWS-NXinet	2100	0200
0100	0119	BBCWS-NX	2-6	World Briefing	NX	BBCWS-NXinet	2100	0200

UTC Time Start	End	Station/ Network	Day(s)	Program Name	Type	Frequncy/ Platform	EDT	Station Time
0100	0129	BBCWS-NX	7	From Our Own Correspondent	NA	BBCWS-NXinet	2100	0200
0100	0129	BBCWS-PR	1	Assignment	ND	Sirius141, NPRfm/am, HD+	2100	0200
0100	0119	BBCWS-PR	2-6	World Briefing	NX	Sirius141, NPRfm/am, HD+	2100	0200
0100	0129	BBCWS-PR	7	From Our Own Correspondent	NA	Sirius141, NPRfm/am, HD+	2100	0200
0100	0259	CBC-R1A	1	Saturday Night Blues	MF	CBAMinet, CBCTinet, CBDinet, CBHAinet, CBlinet, CBNinet, 640, 6160, CBTinet, CBYinet, CBZFinet, CFGBinet	2100	2200
0100	0259	CBC-R1A	2	In the Key of Charles	MV	CBAMinet, CBCTinet, CBDinet, CBHAinet, CBlinet, CBNinet, 640, 6160, CBTinet, CBYinet, CBZFinet, CFGBinet	2100	2200
0100	0159	CBC-R1A	3-7	Q	AC	CBAMinet, CBCTinet, CBDinet, CBHAinet, CBlinet, CBNinet, 640, 6160, CBTinet, CBYinet, CBZFinet, CFGBinet	2100	2200
0100	0159	CBC-R1C	2	Inside the Music	MZ	CBWinet, 990	2100	2000
0100	0144	CBC-R1C	4-7	The Night Time Review	CA	CBWinet, 990	2100	2000
0100	0159	CBC-R1E	1	A Propos	MF	CBCLinet, CBCSinet, CBEinet, 1550, CBLAinet, CBMEinet, CBOinet, CBQTinet, CBVEinet, CFFBinet	2100	2100
0100	0259	CBC-R1E	2	In the Key of Charles	MV	CBCLinet, CBCSinet, CBEinet, 1550, CBLAinet, CBMEinet, CBOinet, CBQTinet, CBVEinet, CFFBinet	2100	2100
0100	0159	CBC-R1E	3-7	Ideas	ED	CBCLinet, CBCSinet, CBEinet, 1550, CBLAinet, CBMEinet, CBOinet, CBQTinet, CBVEinet, CFFBinet	2100	2100
0100	0259	CBC-R1M	1	Vinyl Tap	MP	CBKinet, 540, CBRinet, 1010, CBXinet, 740, CFYKinet, CHAKinet	2100	1900
0100	0129	CBC-R1P	1/2	The World This Weekend	NZ	CBCVinet, CBTKinet, CBUinet, 690, 6160, CBYGinet, CFWHinet	2100	1800
0100	0129	CBC-R1P	3-7	The World At Six	NX	CBCVinet, CBTKinet, CBUinet, 690, 6160, CBYGinet, CFWHinet	2100	1800
0100	0129	CBC-R1S	1/2	The World This Weekend	NZ	Sirius137	2100	2100
0100	0129	CBC-R1S	3-7	The World At Six	NX	Sirius137	2100	2100
0100	0119	CRI-ENG	1/7	News and Reports	NZ	6005, 6020, 6080, 9570, 9580	2100	0900
0100	0126	CRI-ENG	2-6	News and Reports	NZ	6005, 6020, 6080, 9570, 9580	2100	0900
0100	0129	CRI-RTC	1-7	News and Reports	NZ	CRI-RTCinet	2100	0900
0100	0105	DW	1-7	News	NX	DWinet	2100	0300
0100	0104	R.PRG	1/2	News	NX	R.PRGinet, 6200, 7345	2100	0300
0100	0109	R.PRG	3-7	News and Current Affairs	NX	R.PRGinet, 6200, 7345	2100	0300
0100	0159	RNW	1/3	Earthbeat	EV	RNW2inet	2100	0300
0100	0129	RNW	2	Reloaded	VA	RNW2inet	2100	0300
0100	0129	RNW	4	Bridges with Africa	CS	RNW2inet	2100	0300
0100	0129	RNW	5	Curious Orange	DL	RNW2inet	2100	0300
0100	0129	RNW	6	Network Europe Extra	AC	RNW2inet	2100	0300
0100	0129	RNW	7	Radio Books	LD	RNW2inet	2100	0300
0100	0459	RNZ-NAT	2-6	Afternoons with Jim Mora	CA	RNZ-NATinet	2100	1300
0100	0130	RNZI	2-6	BBC World Briefing	NX	RNZIinet, 15720	2100	1300
0100	0159	RTE-R1	1/2	Marian Finucane	CA	RTE-R1inet	2100	0200
0100	0129	RTE-R1	3-7	The Tubridy Show	VA	RTE-R1inet	2100	0200
0100	0459	RTHK-3	1	Sunday Morning	GZ	RTHK-3inet	2100	0900
0100	0459	RTHK-3	7	Saturday Morning with Phil	MP	RTHK-3inet	2100	0900
0100	0129	WRN-NA	1-7	RTE Ireland features	VA	Sirius140, WRN-NAinet	2100	0200
0105	0129	DW	1/2	Newslink	NZ	DWinet	2105	0305
0105	0159	DW	3-7	Newslink Plus	NZ	DWinet	2105	0305
0105	0114	R.PRG	1	Magazine	DL	R.PRGinet, 6200, 7345	2105	0305
0105	0114	R.PRG	2	Mailbox	LI	R.PRGinet, 6200, 7345	2105	0305
0105	0109	RAI-INT	1-6	News Bulletin	NX	RAI-INTinet	2105	0305
0106	0129	RNZ-NAT	1	At the Movies	AC	RNZ-NATinet	2106	1306
0110	0119	R.PRG	3	One on One	PI	R.PRGinet, 6200, 7345	2110	0310
0110	0128	R.PRG	4	Talking Point	DL	R.PRGinet, 6200, 7345	2110	0310
0110	0128	R.PRG	5	Czechs in History (monthly)	AC	R.PRGinet, 6200, 7345	2110	0310
0110	0128	R.PRG	5	Czechs Today (monthly)	CS	R.PRGinet, 6200, 7345	2110	0310
0110	0128	R.PRG	5	Spotlight (fortnightly)	TR	R.PRGinet, 6200, 7345	2110	0310
0110	0118	R.PRG	6	Panorama	CS	R.PRGinet, 6200, 7345	2110	0310
0110	0118	R.PRG	7	Business News	BE	R.PRGinet, 6200, 7345	2110	0310
0115	0118	R.PRG	1	Sound Czech	LL	R.PRGinet, 6200, 7345	2119	0319
0115	0119	R.PRG	2	Letter from Prague	DL	R.PRGinet, 6200, 7345	2115	0315
0119	0128	R.PRG	1	One on One	PI	R.PRGinet, 6200, 7345	2119	0319
0119	0128	R.PRG	2	Czech Books (fortnightly)	AC	R.PRGinet, 6200, 7345	2119	0319
0119	0128	R.PRG	2	Magic Carpet (monthly)	MZ	R.PRGinet, 6200, 7345	2119	0319
0119	0128	R.PRG	2	Music Profile (monthly)	MX	R.PRGinet, 6200, 7345	2119	0319
0119	0128	R.PRG	7	The Arts	AC	R.PRGinet, 6200, 7345	2119	0319
0120	0129	BBCWS-AM	2-6	World Business Report	BE	XM131	2120	0220
0120	0129	BBCWS-IE	2-6	World Business Report	BE	BBCWS-IEinet	2120	0220
0120	0129	BBCWS-NX	2-6	World Business Report	BE	BBCWS-NXinet	2120	0220
0120	0129	BBCWS-PR	2-6	World Business Report	BE	Sirius141, NPRfm/am, HD+	2120	0220
0120	0126	CRI-ENG	1	Reports from Developing Countries	BE	6005, 6020, 6080, 9570, 9580	2120	0900
0120	0126	CRI-ENG	7	CRI Roundup	NX	6005, 6020, 6080, 9570, 9580	2120	0900
0120	0128	R.PRG	3	Sports News	SP	R.PRGinet, 6200, 7345	2120	0320
0127	0151	CRI-ENG	1	In the Spotlight	DL	6005, 6020, 6080, 9570, 9580	2127	0927
0127	0151	CRI-ENG	2	People in the Know	GL	6005, 6020, 6080, 9570, 9580	2127	0927
0127	0151	CRI-ENG	3	Biz China	BE	6005, 6020, 6080, 9570, 9580	2127	0927
0127	0151	CRI-ENG	4	China Horizons	DL	6005, 6020, 6080, 9570, 9580	2127	0927
0127	0151	CRI-ENG	5	Voices from Other Lands	DL	6005, 6020, 6080, 9570, 9580	2127	0927
0127	0151	CRI-ENG	6	Life in China	DL	6005, 6020, 6080, 9570, 9580	2127	0927
0127	0151	CRI-ENG	7	Listener's Garden	LI	6005, 6020, 6080, 9570, 9580	2127	0927

UTC Time Start	End	Station/ Network	Day(s)	Program Name	Type	Frequency/ Platform	Station EDT	Time
0130	0159	ABC-RA	1	Talking Point	CA	RAinet	2130	1130
0130	0159	ABC-RN	1	MovieTime	AC	ABC-RNinet	2130	1130
0130	0139	BBCWS-AM	2	The Instant Guide	NA	XM131	2130	0230
0130	0139	BBCWS-AM	3-6	World Briefing	NX	XM131	2130	0230
0130	0159	BBCWS-AM	7	World Football	SP	XM131	2130	0230
0130	0159	BBCWS-IE	1	Charlie Gillett's World of Music	MZ	BBCWS-IEinet	2130	0230
0130	0159	BBCWS-IE	2	Science in Action	ST	BBCWS-IEinet	2130	0230
0130	0159	BBCWS-IE	3-7	The Strand	AC	BBCWS-IEinet	2130	0230
0130	0159	BBCWS-NX	1	Reporting Religion	CS	BBCWS-NXinet	2130	0230
0130	0139	BBCWS-NX	2-6	World Briefing	NX	BBCWS-NXinet	2130	0230
0130	0159	BBCWS-NX	7	Global Business	BE	BBCWS-NXinet	2130	0230
0130	0159	BBCWS-PR	1	Charlie Gillett's World of Music	MZ	Sirius141, NPRfm/am, HD+	2130	0230
0130	0139	BBCWS-PR	2-6	World Briefing	NX	Sirius141, NPRfm/am, HD+	2130	0230
0130	0159	BBCWS-PR	7	Global Business	BE	Sirius141, NPRfm/am, HD+	2130	0230
0130	0159	CBC-R1C	2	The Debaters	LE	CBWinet, 990	2130	2030
0130	0159	CBC-R1M	2	C'est la vie	DL	CBKinet, 540, CBRinet, 1010, CBXinet, 740, CFYKinet, CHAKinet	2130	1930
0130	0229	CBC-R1M	3	Dispatches	NA	CBKinet, 540, CBRinet, 1010, CBXinet, 740, CFYKinet, CHAKinet	2130	1930
0130	0159	CBC-R1P	1	Laugh Out Loud	LE	CBCVinet, CBTKinet, CBUinet, 690, 6160, CBYGinet, CFWHinet	2130	1830
0130	0229	CBC-R1P	2	Dispatches	NA	CBCVinet, CBTKinet, CBUinet, 690, 6160, CBYGinet, CFWHinet	2130	1830
0130	0229	CBC-R1P	3	As It Happens	NZ	CBCVinet, CBTKinet, CBUinet, 690, 6160, CBYGinet, CFWHinet	2130	1830
0130	0259	CBC-R1P	4-7	As It Happens	NZ	CBCVinet, CBTKinet, CBUinet, 690, 6160, CBYGinet, CFWHinet	2130	1830
0130	0159	CBC-R1S	1	The Age of Persuasion	CS	Sirius137	2130	2130
0130	0159	CBC-R1S	2	WireTap	LE	Sirius137	2130	2130
0130	0229	CBC-R1S	3	As It Happens	NA	Sirius137	2130	2130
0130	0259	CBC-R1S	4-7	As It Happens	NA	Sirius137	2130	2130
0130	0159	CRI-RTC	1	China Horizons	DL	CRI-RTCinet	2130	0930
0130	0159	CRI-RTC	2	Frontline	GL	CRI-RTCinet	2130	0930
0130	0159	CRI-RTC	3	Biz China	BE	CRI-RTCinet	2130	0930
0130	0159	CRI-RTC	4	In the Spotlight	DL	CRI-RTCinet	2130	0930
0130	0159	CRI-RTC	5	Voices from Other Lands	DL	CRI-RTCinet	2130	0930
0130	0159	CRI-RTC	6	Life in China	DL	CRI-RTCinet	2130	0930
0130	0159	CRI-RTC	7	Listeners' Garden	LI	CRI-RTCinet	2130	0930
0130	0144	DW	1/2	Sports Report	SP	DWinet	2130	0330
0130	0229	IRIB	1-7	Koran Reading, News and features	VA	IRIBinet, 6120, 7160	2130	0500
0130	0159	RNW	1/3	Curious Orange	DL	RNW2inet	2130	0330
0130	0159	RNW	2/7	Bridges with Africa	CS	RNW2inet	2130	0330
0130	0159	RNW	4	The State We're In Midweek Edition	CS	RNW2inet	2130	0330
0130	0159	RNW	5	Radio Books	LD	RNW2inet	2130	0330
0130	0159	RNW	6	Earthbeat	EV	RNW2inet	2130	0330
0130	0137	RNZI	2-6	Pacific Regional News	NX	RNZIinet, 15720	2130	1330
0130	0229	RTE-R1	3-7	Today with Pat Kenny	CA	RTE-R1inet	2130	0230
0130	0459	RTHK-3	2-6	Morning Brew	GZ	RTHK-3inet	2130	0930
0130	0159	WRN-NA	1-7	Radio Sweden features	VA	Sirius140, WRN-NAinet	2130	0230
0135	0159	ABC-RA	7	Rear Vision	NA	RAinet	2135	1135
0138	0159	RNZI	2-6	Dateline Pacific	NA	RNZIinet, 15720	2138	1338
0140	0159	BBCWS-AM	2	Over to You	LI	XM131	2140	0240
0140	0149	BBCWS-AM	3-6	Analysis	NA	XM131	2140	0240
0140	0149	BBCWS-NX	2-6	Analysis	NA	BBCWS-NXinet	2140	0240
0140	0159	BBCWS-PR	2-6	Business Daily	BE	Sirius141, NPRfm/am, HD+	2140	0240
0145	0159	CBC-R1C	4-7	Outfront	DL	CBWinet, 990	2145	2045
0145	0159	DW	1	Inspired Minds	AC	DWinet	2145	0345
0145	0159	DW	2	Radio D	LL	DWinet	2145	0345
0150	0159	BBCWS-AM	3-6	Sports Roundup	SP	XM131	2150	0250
0150	0159	BBCWS-NX	2-6	Sports Roundup	SP	BBCWS-NXinet	2150	0250
0152	0157	CRI-ENG	1-7	Learning Chinese Now	LL	6005, 6020, 6080, 9570, 9580	2152	0952
0200	0229	ABC-RA	1	Artworks	AC	RAinet	2200	1200
0200	0259	ABC-RA	2-6	The World Today	NZ	RAinet	2200	1200
0200	0229	ABC-RA	7	Asia Review	CA	RAinet	2200	1200
0200	0259	ABC-RN	1	The National Interest	CA	ABC-RNinet	2200	1200
0200	0259	ABC-RN	2-6	The World Today	NZ	ABC-RNinet	2200	1200
0200	0259	ABC-RN	7	The Science Show	ST	ABC-RNinet	2200	1200
0200	0229	BBCWS-AM	1/7	The World Today	NZ	XM131	2200	0300
0200	0259	BBCWS-AM	2-6	The World Today	NZ	XM131	2200	0300
0200	0229	BBCWS-IE	1-7	The World Today	NZ	BBCWS-IEinet	2200	0300
0200	0229	BBCWS-NX	1/7	The World Today	NZ	BBCWS-NXinet	2200	0300
0200	0359	BBCWS-NX	2-6	The World Today	NZ	BBCWS-NXinet	2200	0300
0200	0229	BBCWS-PR	1/7	The World Today	NZ	Sirius141, NPRfm/am, HD+	2200	0300
0200	0359	BBCWS-PR	2-6	The World Today	NZ	Sirius141, NPRfm/am, HD+	2200	0300
0200	0259	CBC-R1A	3	Quirks and Quarks	ST	CBAMinet, CBCTinet, CBDinet, CBHAinet, CBIinet, CBNinet, 640, 6160, CBTinet, CBYinet, CBZFinet, CFGBinet	2200	2300
0200	0259	CBC-R1A	4	The Vinyl Cafe	LE	CBAMinet, CBCTinet, CBDinet, CBHAinet, CBIinet, CBNinet, 640, 6160, CBTinet, CBYinet, CBZFinet, CFGBinet	2200	2300

UTC Time Start	End	Station/ Network	Day(s)	Program Name	Type	Frequency/ Platform	EDT	Station Time
0200	0229	CBC-R1A	5	Afghanada	LD	CBAMinet, CBCTinet, CBDinet, CBHAinet, CBlinet, CBNinet, 640, 6160, CBTinet, CBYinet, CBZFinet, CFGBinet	2200	2300
0200	0259	CBC-R1A	6	Writers and Company	AC	CBAMinet, CBCTinet, CBDinet, CBHAinet, CBlinet, CBNinet, 640, 6160, CBTinet, CBYinet, CBZFinet, CFGBinet	2200	2300
0200	0359	CBC-R1A	7	Vinyl Tap	MP	CBAMinet, CBCTinet, CBDinet, CBHAinet, CBlinet, CBNinet, 640, 6160, CBTinet, CBYinet, CBZFinet, CFGBinet	2200	2300
0200	0259	CBC-R1C	1	A Propos	MF	CBWinet, 990	2200	2100
0200	0359	CBC-R1C	2	In the Key of Charles	MV	CBWinet, 990	2200	2100
0200	0259	CBC-R1C	3-7	Ideas	ED	CBWinet, 990	2200	2100
0200	0359	CBC-R1E	1	Saturday Night Blues	MF	CBCLinet, CBCSinet, CBEinet, 1550, CBLAinet, CBMEinet, CBOinet, CBQTinet, CBVEinet, CFFBinet	2200	2200
0200	0259	CBC-R1E	3-7	Q	AC	CBCLinet, CBCSinet, CBEinet, 1550, CBLAinet, CBMEinet, CBOinet, CBQTinet, CBVEinet, CFFBinet	2200	2200
0200	0259	CBC-R1M	2	Inside the Music	MZ	CBKinet, 540, CBRinet, 1010, CBXinet, 740, CFYKinet, CHAKinet	2200	2000
0200	0244	CBC-R1M	4-7	The Night Time Review	CA	CBKinet, 540, CBRinet, 1010, CBXinet, 740, CFYKinet, CHAKinet	2200	2000
0200	0359	CBC-R1P	1	Vinyl Tap	MP	CBCVinet, CBTKinet, CBUinet, 690, 6160, CBYGinet, CFWHinet	2200	1900
0200	0359	CBC-R1S	1	Definitely Not the Opera	CS	Sirius137	2200	2200
0200	0259	CBC-R1S	2	Deep Roots	MF	Sirius137	2200	2200
0200	0229	CRI-RTC	1-7	News and Reports	NZ	CRI-RTCinet	2200	1000
0200	0259	CRI-WASH	2/3	China Beat	MP	CRI-WASHinet	2200	1000
0200	0219	CRI-WASH	4-1	People in the Know	GL	CRI-WASHinet	2200	1000
0200	0205	DW	1-7	News	NX	DWinet	2200	0400
0200	0204	R.PRG	1/2	News	NX	R.PRGinet, 6200, 7345	2200	0400
0200	0209	R.PRG	3-7	News and Current Affairs	NX	R.PRGinet, 6200, 7345	2200	0400
0200	0259	RAE	3-7	News, Reports, Features, Tangos	NZ	11710, RAEinet	2200	2300
0200	0229	RNW	1	The State We're In	CS	RNW2inet	2200	0400
0200	0229	RNW	2/6	Network Europe Extra	AC	RNW2inet	2200	0400
0200	0229	RNW	3	Earthbeat	EV	RNW2inet	2200	0400
0200	0229	RNW	4	Bridges with Africa	CS	RNW2inet	2200	0400
0200	0229	RNW	5	Curious Orange	DL	RNW2inet	2200	0400
0200	0229	RNW	7	Radio Books	LD	RNW2inet	2200	0400
0200	0529	RNZ-NAT	7	Music 101 with Kirsten Johnstone	MR	RNZ-NATinet	2200	1400
0200	0259	RNZI	2-6	(see RNZ-NAT)		RNZIinet, 15720	2200	1400
0200	0259	RTE-R1	1	Saturdayview	GL	RTE-R1inet	2200	0300
0200	0259	RTE-R1	2	This Week	NA	RTE-R1inet	2200	0300
0200	0209	RTI	1-7	News	NX	5950	2200	1100
0200	0354	SABC-SAFM	1/7	Early Classics	MC	SABC-SAFMinet	2200	0400
0200	0359	SABC-SAFM	2-6	Heads Up	GZ	SABC-SAFMinet	2200	0400
0200	0229	VOR-WS	1/2	Moscow Mailbag	LI	6100, 6240, 7250, 12040, 13735	2200	0500
0200	0229	VOR-WS	3-7	Russia and the World	NA	6100, 6240, 7250, 12040, 13735	2200	0500
0200	0229	WRN-NA	1-7	Radio Prague features	VA	Sirius140, WRN-NAinet	2200	0300
0204	0224	RNZ-NAT	1	Comedy	LE	RNZ-NATinet	2204	1404
0205	0229	DW	1/2	Newslink	NZ	DWinet	2205	0405
0205	0259	DW	3-7	Newslink Plus	NZ	DWinet	2205	0405
0205	0214	R.PRG	1	Magazine	DL	R.PRGinet, 6200, 7345	2205	0405
0205	0214	R.PRG	2	Mailbox	LI	R.PRGinet, 6200, 7345	2205	0405
0205	0209	RAI-INT	1-6	News Bulletin	NX	RAI-INTinet	2205	0405
0210	0219	R.PRG	3	One on One	PI	R.PRGinet, 6200, 7345	2210	0410
0210	0228	R.PRG	4	Talking Point	DL	R.PRGinet, 6200, 7345	2210	0410
0210	0228	R.PRG	5	Czechs in History (monthly)	AC	R.PRGinet, 6200, 7345	2210	0410
0210	0228	R.PRG	5	Czechs Today (monthly)	CS	R.PRGinet, 6200, 7345	2210	0410
0210	0228	R.PRG	5	Spotlight (fortnightly)	TR	R.PRGinet, 6200, 7345	2210	0410
0210	0218	R.PRG	6	Panorama	CS	R.PRGinet, 6200, 7345	2210	0410
0210	0218	R.PRG	7	Business News	BE	R.PRGinet, 6200, 7345	2210	0410
0210	0220	RTI	1	Time Traveler	AC	5950	2210	1110
0210	0220	RTI	2	Women Making Waves	CS	5950	2210	1110
0210	0255	RTI	3	We've Got Mail!	LI	5950	2210	1110
0210	0255	RTI	4	Health Beats	HM	5950	2210	1110
0210	0255	RTI	5	Ilha Formosa	DL	5950	2210	1110
0210	0255	RTI	6	News Talk	NA	5950	2210	1110
0210	0255	RTI	7	The Occidental Tourist	TR	5950	2210	1110
0215	0218	R.PRG	1	Sound Czech	LL	R.PRGinet, 6200, 7345	2215	0415
0215	0219	R.PRG	2	Letter from Prague	DL	R.PRGinet, 6200, 7345	2215	0415
0219	0228	R.PRG	1	One on One	PI	R.PRGinet, 6200, 7345	2219	0419
0219	0228	R.PRG	2	Czech Books (fortnightly)	AC	R.PRGinet, 6200, 7345	2219	0419
0219	0228	R.PRG	2	Magic Carpet (monthly)	MZ	R.PRGinet, 6200, 7345	2219	0419
0219	0228	R.PRG	2	Music Profile (monthly)	MX	R.PRGinet, 6200, 7345	2219	0419
0219	0228	R.PRG	7	The Arts	AC	R.PRGinet, 6200, 7345	2219	0419
0220	0239	CRI-WASH	4-1	China Horizons	DL	CRI-WASHinet	2220	1020
0220	0228	R.PRG	3	Sports News	SP	R.PRGinet, 6200, 7345	2220	0420
0220	0240	RTI	1	Spotlight	DL	5950	2220	1120
0220	0230	RTI	2	Chinese to Go	LL	5950	2220	1120
0220	0235	RTI	4	People	DL	5950	2220	1120
0220	0235	RTI	5	Breakfast Club	ED	5950	2220	1120
0220	0240	RTI	6	Taiwan Indie	AC	5950	2220	1120

UTC Time Start	End	Station/ Network	Day(s)	Program Name	Type	Frequency/ Platform	Station EDT	Time
0220	0255	RTI	7	Groove Zone	MP	5950	2220	1120
0225	0255	RNZ-NAT	1	Books	AC	RNZ-NATinet	2225	1425
0230	0259	ABC-RA	1	Australian Country Style	MW	RAinet	2230	1230
0230	0259	ABC-RA	7	All in the Mind	HM	RAinet	2230	1230
0230	0259	BBCWS-AM	1	The Interview	PI	XM131	2230	0330
0230	0259	BBCWS-AM	7	Politics UK	GL	XM131	2230	0330
0230	0259	BBCWS-IE	1	Discovery	ST	BBCWS-IEinet	2230	0330
0230	0239	BBCWS-IE	2	The Instant Guide	NA	BBCWS-IEinet	2230	0330
0230	0259	BBCWS-IE	3-7	Outlook	GZ	BBCWS-IEinet	2230	0330
0230	0259	BBCWS-NX	1	The Interview	PI	BBCWS-NXinet	2230	0330
0230	0259	BBCWS-NX	7	Politics UK	GL	BBCWS-NXinet	2230	0330
0230	0259	BBCWS-PR	1	Business Weekly	BE	Sirius141, NPRfm/am, HD+	2230	0330
0230	0259	BBCWS-PR	7	One Planet	EV	Sirius141, NPRfm/am, HD+	2230	0330
0230	0259	CBC-R1A	5	WireTap	LE	CBAMinet, CBCTinet, CBDinet, CBHAinet, CBIinet, CBNinet, 640, 6160, CBTinet, CBYinet, CBZFinet, CFGBinet	2230	2330
0230	0259	CBC-R1M	2	The Debaters	LE	CBKinet, 540, CBRinet, 1010, CBXinet, 740, CFYKinet, CHAKinet	2230	2030
0230	0259	CBC-R1P	2	C'est la vie	DL	CBCVinet, CBTKinet, CBUinet, 690, 6160, CBYGinet, CFWHinet	2230	1930
0230	0329	CBC-R1P	3	Dispatches	NA	CBCVinet, CBTKinet, CBUinet, 690, 6160, CBYGinet, CFWHinet	2230	1930
0230	0259	CBC-R1S	2	The Choice	GI	Sirius137	2230	2230
0230	0259	CRI-RTC	1	China Horizons	DL	CRI-RTCinet	2230	1030
0230	0259	CRI-RTC	2	Frontline	GL	CRI-RTCinet	2230	1030
0230	0259	CRI-RTC	3	Biz China	BE	CRI-RTCinet	2230	1030
0230	0259	CRI-RTC	4	In the Spotlight	DL	CRI-RTCinet	2230	1030
0230	0259	CRI-RTC	5	Voices from Other Lands	DL	CRI-RTCinet	2230	1030
0230	0259	CRI-RTC	6	Life in China	DL	CRI-RTCinet	2230	1030
0230	0259	CRI-RTC	7	Listeners' Garden	LI	CRI-RTCinet	2230	1030
0230	0259	DW	1	Insight	NA	DWinet	2230	0430
0230	0259	DW	2	A World of Music	M	DWinet	2230	0430
0230	0239	KBS-WR	1-7	News	NX	9560	2230	1130
0230	0259	RNW	2/5	Radio Books	LD	RNW2inet	2230	0430
0230	0259	RNW	3	Curious Orange	DL	RNW2inet	2230	0430
0230	0259	RNW	4	The State We're In Midweek Edition	CS	RNW2inet	2230	0430
0230	0259	RNW	6	Earthbeat	EV	RNW2inet	2230	0430
0230	0259	RNW	7	Bridges with Africa	CS	RNW2inet	2230	0430
0230	0259	RTE-R1	3-7	Liveline	LI	RTE-R1inet	2230	0330
0230	0255	RTI	2	Soundwaves	MP	5950	2230	1130
0230	0259	VOR-WS	2	Timelines	DL	6100, 6240, 7250, 12040, 13735	2230	0530
0230	0259	VOR-WS	3/6	Kaleidoscope	CS	6100, 6240, 7250, 12040, 13735	2230	0530
0230	0259	VOR-WS	4	Russian by Radio	LL	6100, 6240, 7250, 12040, 13735	2230	0530
0230	0259	VOR-WS	5	The VOR Treasure Store	AC	6100, 6240, 7250, 12040, 13735	2230	0530
0230	0259	VOR-WS	6	The Christian Message from Moscow	CS	6100, 6240, 7250, 12040, 13735	2230	0530
0230	0259	VOR-WS	7	Moscow Yesterday and Today	AC	6100, 6240, 7250, 12040, 13735	2230	0530
0230	0259	WRN-NA	1-7	KBS World Radio features	VA	Sirius140, WRN-NAinet	2230	0330
0235	0255	RTI	4	Jade Bells and Bamboo Pipes	MF	5950	2235	1135
0235	0250	RTI	5	Instant Noodles	LE	5950	2235	1135
0240	0259	BBCWS-IE	2	Over to You	LI	BBCWS-IEinet	2240	0340
0240	0259	CRI-WASH	4-1	Cultural Horizons	AC	CRI-WASHinet	2240	1040
0240	0259	KBS-WR	1	Korean Pop Interactive	MP	9560	2240	1140
0240	0244	KBS-WR	2-6	News Commentary	NC	9560	2240	1140
0240	0259	KBS-WR	7	Worldwide Friendship	LI	9560	2240	1140
0240	0255	RTI	1	On the Line	PI	5950	2240	1140
0240	0255	RTI	6	Taiwan Outlook	BE	5950	2240	1140
0245	0259	CBC-R1M	4-7	Outfront	DL	CBKinet, 540, CBRinet, 1010, CBXinet, 740, CFYKinet, CHAKinet	2245	2045
0245	0259	KBS-WR	2	Faces of Korea	CS	9560	2245	1145
0245	0259	KBS-WR	3	Business Watch	BE	9560	2245	1145
0245	0259	KBS-WR	4	Culture on the Move	AC	9560	2245	1145
0245	0259	KBS-WR	5	Korea Today and Tomorrow	DL	9560	2245	1145
0245	0259	KBS-WR	6	Seoul Report	DL	9560	2245	1145
0250	0255	RTI	5	Chinese to Go	LL	5950	2250	1150
0300	0329	ABC-RA	1	The Ark (til 1/09)	CS	RAinet	2300	1300
0300	0329	ABC-RA	1	Sunday Profile (from 2/09)	NA	RAinet	2300	1300
0300	0309	ABC-RA	2-6	Sports Report	SP	RAinet	2300	1300
0300	0329	ABC-RA	7	Pacific Review	NZ	RAinet	2300	1300
0300	0329	ABC-RN	1	Rear Vision (til 1/09)	NA	ABC-RNinet	2300	1300
0300	0329	ABC-RN	1	Sunday Profile (from 2/09)	NA	ABC-RNinet	2300	1300
0300	0334	ABC-RN	2/7	All in the Mind	HM	ABC-RNinet	2300	1300
0300	0359	ABC-RN	3	The Spirit of Things	CS	ABC-RNinet	2300	1300
0300	0359	ABC-RN	4	Feature program (from 2/09)	ND	ABC-RNinet	2300	1300
0300	0359	ABC-RN	4	Radio Eye (til 1/09)	ND	ABC-RNinet	2300	1300
0300	0359	ABC-RN	5	Hindsight	AC	ABC-RNinet	2300	1300
0300	0359	ABC-RN	6	Counterpoint	CA	ABC-RNinet	2300	1300
0300	0329	BBCWS-AM	1	Documentary feature or series	ND	XM131	2300	0400
0300	0329	BBCWS-AM	2	Global Business	BE	XM131	2300	0400
0300	0329	BBCWS-AM	3-7	Outlook	GZ	XM131	2300	0400
0300	0329	BBCWS-IE	1/7	The World Today	NZ	BBCWS-IEinet	2300	0400
0300	0359	BBCWS-IE	2-6	The World Today	NZ	BBCWS-IEinet	2300	0400

UTC Time Start	End	Station/ Network	Day(s)	Program Name	Type	Frequency/ Platform	EDT	Station Time
0300	0329	BBCWS-NX	1/7	The World Today	NZ	BBCWS-NXinet	2300	0400
0300	0329	BBCWS-PR	1/7	The World Today	NZ	Sirius141, NPRfm/am, HD+	2300	0400
0300	0359	CBC-R1A	1	Global Arts and Entertainment	AC	CBAMinet, CBCTinet, CBDinet, CBHAinet, CBlinet, CBNinet, 640, 6160, CBTinet, CBYinet, CBZFinet, CFGBinet	2300	0000
0300	0359	CBC-R1A	2	Tonic	MJ	CBAMinet, CBCTinet, CBDinet, CBHAinet, CBlinet, CBNinet, 640, 6160, CBTinet, CBYinet, CBZFinet, CFGBinet	2300	0000
0300	0329	CBC-R1A	3	From Our Own Correspondent	NA	CBAMinet, CBCTinet, CBDinet, CBHAinet, CBlinet, CBNinet, 640, 6160, CBTinet, CBYinet, CBZFinet, CFGBinet	2300	0000
0300	0359	CBC-R1A	4	The Choice	GI	CBAMinet, CBCTinet, CBDinet, CBHAinet, CBlinet, CBNinet, 640, 6160, CBTinet, CBYinet, CBZFinet, CFGBinet	2300	0000
0300	0359	CBC-R1A	5	And the Winner Is	FD	CBAMinet, CBCTinet, CBDinet, CBHAinet, CBlinet, CBNinet, 640, 6160, CBTinet, CBYinet, CBZFinet, CFGBinet	2300	0000
0300	0359	CBC-R1A	6	Rewind	GI	CBAMinet, CBCTinet, CBDinet, CBHAinet, CBlinet, CBNinet, 640, 6160, CBTinet, CBYinet, CBZFinet, CFGBinet	2300	0000
0300	0459	CBC-R1C	1	Saturday Night Blues	MF	CBWinet, 990	2300	2200
0300	0359	CBC-R1C	3-7	Q	AC	CBWinet, 990	2300	2200
0300	0459	CBC-R1E	2	Tonic	MJ	CBCLinet, CBCSinet, CBEinet, 1550, CBLAinet, CBMEinet, CBOinet, CBQTinet, CBVEinet, CFFBinet	2300	2300
0300	0359	CBC-R1E	3	Quirks and Quarks	ST	CBCLinet, CBCSinet, CBEinet, 1550, CBLAinet, CBMEinet, CBOinet, CBQTinet, CBVEinet, CFFBinet	2300	2300
0300	0359	CBC-R1E	4	The Vinyl Cafe	LE	CBCLinet, CBCSinet, CBEinet, 1550, CBLAinet, CBMEinet, CBOinet, CBQTinet, CBVEinet, CFFBinet	2300	2300
0300	0329	CBC-R1E	5	Afghanada	LD	CBCLinet, CBCSinet, CBEinet, 1550, CBLAinet, CBMEinet, CBOinet, CBQTinet, CBVEinet, CFFBinet	2300	2300
0300	0359	CBC-R1E	6	Writers and Company	AC	CBCLinet, CBCSinet, CBEinet, 1550, CBLAinet, CBMEinet, CBOinet, CBQTinet, CBVEinet, CFFBinet	2300	2300
0300	0459	CBC-R1E	7	Vinyl Tap	MP	CBCLinet, CBCSinet, CBEinet, 1550, CBLAinet, CBMEinet, CBOinet, CBQTinet, CBVEinet, CFFBinet	2300	2300
0300	0359	CBC-R1M	1	A Propos	MF	CBKinet, 540, CBRinet, 1010, CBXinet, 740, CFYKinet, CHAKinet	2300	2100
0300	0459	CBC-R1M	2	In the Key of Charles	MV	CBKinet, 540, CBRinet, 1010, CBXinet, 740, CFYKinet, CHAKinet	2300	2100
0300	0359	CBC-R1M	3-7	Ideas	ED	CBKinet, 540, CBRinet, 1010, CBXinet, 740, CFYKinet, CHAKinet	2300	2100
0300	0359	CBC-R1P	2	Inside the Music	MZ	CBCVinet, CBTKinet, CBUinet, 690, 6160, CBYGinet, CFWHinet	2300	2000
0300	0344	CBC-R1P	4-7	The Night Time Review	CA	CBCVinet, CBTKinet, CBUinet, 690, 6160, CBYGinet, CFWHinet	2300	2000
0300	0459	CBC-R1S	1	Vinyl Tap	MP	Sirius137	2300	2300
0300	0359	CBC-R1S	3-7	Q	AC	Sirius137	2300	2300
0300	0319	CRI-ENG	1/7	News and Reports	NZ	6190, 9690, 9790	2300	1100
0300	0326	CRI-ENG	2-6	News and Reports	NZ	6190, 9690, 9790	2300	1100
0300	0329	CRI-RTC	1-7	News and Reports	NZ	CRI-RTCinet	2300	1100
0300	0559	CRI-WASH	1-7	Feature Programs	VA	CRI-WASHinet	2300	1100
0300	0305	DW	1-7	News	NX	DWinet	2300	0500
0300	0309	KBS-WR	1-7	News	NX	KBS-WR1inet	2300	1200
0300	0329	RNW	1	Network Europe Week	NA	RNW2inet	2300	0500
0300	0329	RNW	2	Network Europe Extra	AC	RNW2inet	2300	0500
0300	0329	RNW	3-7	Newsline	NZ	RNW2inet	2300	0500
0300	0307	RNZI	2-7	Pacific Regional News	NX	RNZIinet, 15720	2300	1500
0300	0359	RTE-R1	1	The Late Session	MF	RTE-R1inet	2300	0400
0300	0329	RTE-R1	2	This Business	BE	RTE-R1inet	2300	0400
0300	0359	RTE-R1	3-7	The Radio 1 Music Collection	MF	RTE-R1inet	2300	0400
0300	0314	SABC-CHAF	1-7	News	NX	CHAFStudio1inet	2300	0500
0300	0319	VOR-WS	2	Sunday Panorama	NA	6100, 6155, 6240, 7250, 12040, 13735, VOR2inet	2300	0600
0300	0329	VOR-WS	3-1	News and Views	NA	6100, 6155, 6240, 7250, 12040, 13735, VOR2inet	2300	0600
0300	0329	WRN-NA	1-7	Polish Radio External Service features	VA	Sirius140, WRN-NAinet	2300	0400
0304	0359	RNZ-NAT	1	The Sunday Drama	LD	RNZ-NATinet	2304	1504
0305	0324	DW	1	In-Box	MB	DWinet	2305	0505
0305	0329	DW	2	Newslink	NZ	DWinet	2305	0505
0305	0359	DW	3-7	Newslink Plus	NZ	DWinet	2305	0505
0308	0329	RNZI	2-7	Dateline Pacific	NA	RNZIinet, 15720	2308	1508
0310	0459	ABC-RA	2-6	In the Loop	CA	RAinet	2310	1310
0310	0314	KBS-WR	2-6	News Commentary	NC	KBS-WR1inet	2310	1210
0315	0359	KBS-WR	1	Korean Pop Interactive	MP	KBS-WR1inet	2315	1215
0315	0344	KBS-WR	2-6	Seoul Calling	NZ	KBS-WR1inet	2315	1215
0315	0359	KBS-WR	7	Worldwide Friendship	LI	KBS-WR1inet	2315	1215
0315	0319	RAI-INT	2-6	News Bulletin	NX	RAI-INTinet	2305	0505
0315	0359	SABC-CHAF	1	Straight Talk	NA	CHAFStudio1inet	2315	0515
0315	0359	SABC-CHAF	2	African Music	MF	CHAFStudio1inet	2315	0515
0315	0359	SABC-CHAF	3	The Inner Voice	CS	CHAFStudio1inet	2315	0515
0315	0359	SABC-CHAF	4	Tam Tam Express	GL	CHAFStudio1inet	2315	0515
0315	0359	SABC-CHAF	5	Our Heritage	AC	CHAFStudio1inet	2315	0515
0315	0359	SABC-CHAF	6	Gateway to Africa	BE	CHAFStudio1inet	2315	0515
0315	0359	SABC-CHAF	7	Sport	SP	CHAFStudio1inet	2315	0515
0320	0326	CRI-ENG	1	Reports from Developing Countries	BE	6190, 9690, 9790	2320	1120
0320	0326	CRI-ENG	7	CRI Roundup	NX	6190, 9690, 9790	2320	1120
0320	0324	VOR-WS	2	A Stroll Around the Kremlin	AC	6100, 6155, 6240, 7250, 12040, 13735, VOR2inet	2320	0620
0325	0329	DW	1	Mission Europe	LL	DWinet	2325	0525
0325	0329	VOR-WS	2	Legends of Russian Sports	SP	6100, 6155, 6240, 7250, 12040, 13735, VOR2inet	2325	0625
0327	0351	CRI-ENG	1	In the Spotlight	DL	6190, 9690, 9790	2327	1127
0327	0351	CRI-ENG	2	People in the Know	GL	6190, 9690, 9790	2327	1127

UTC Time Start	End	Station/ Network	Day(s)	Program Name	Type	Frequncy/ Platform	Station EDT	Time
0327	0351	CRI-ENG	3	Biz China	BE	6190, 9690, 9790	2327	1127
0327	0351	CRI-ENG	4	China Horizons	DL	6190, 9690, 9790	2327	1127
0327	0351	CRI-ENG	5	Voices from Other Lands	DL	6190, 9690, 9790	2327	1127
0327	0351	CRI-ENG	6	Life in China	DL	6190, 9690, 9790	2327	1127
0327	0351	CRI-ENG	7	Listener's Garden	LI	6190, 9690, 9790	2327	1127
0330	0359	ABC-RA	1	MovieTime	AC	RAinet	2330	1330
0330	0344	ABC-RA	7	Asia Pacific Business	BE	RAinet	2330	1330
0330	0359	ABC-RN	1	Rear Vision (from 2/09)	NA	ABC-RNinet	2330	1330
0330	0359	ABC-RN	1	Street Stories (til 1/09)	DL	ABC-RNinet	2330	1330
0330	0359	BBCWS-AM	1	The Interview	PI	XM131	2330	0430
0330	0359	BBCWS-AM	2	Charlie Gillett's World of Music	MZ	XM131	2330	0430
0330	0359	BBCWS-AM	3	Health Check	HM	XM131	2330	0430
0330	0359	BBCWS-AM	4	Digital Planet	DX	XM131	2330	0430
0330	0359	BBCWS-AM	5	Discovery	ST	XM131	2330	0430
0330	0359	BBCWS-AM	6	One Planet	EV	XM131	2330	0430
0330	0359	BBCWS-AM	7	Science in Action	ST	XM131	2330	0430
0330	0359	BBCWS-IE	1	The Interview	PI	BBCWS-IEinet	2330	0430
0330	0359	BBCWS-IE	7	World Football	SP	BBCWS-IEinet	2330	0430
0330	0359	BBCWS-NX	1	Politics UK	GL	BBCWS-NXinet	2330	0430
0330	0359	BBCWS-NX	7	The Interview	PI	BBCWS-NXinet	2330	0430
0330	0359	BBCWS-PR	1	Reporting Religion	CS	Sirius141, NPRfm/am, HD+	2330	0430
0330	0359	BBCWS-PR	7	The Interview	PI	Sirius141, NPRfm/am, HD+	2330	0430
0330	0359	CBC-R1A	3	Culture Shock	CS	CBAMinet, CBCTinet, CBDinet, CBHAinet, CBIinet, CBNinet, 640, 6160, CBTinet, CBYinet, CBZFinet, CFGBinet	2330	0030
0330	0359	CBC-R1E	5	WireTap	LE	CBCLinet, CBCSinet, CBEinet, 1550, CBLAinet, CBMEinet, CBOinet, CBQTinet, CBVEinet, CFFBinet	2330	2330
0330	0359	CBC-R1P	2	The Debaters	LE	CBCVinet, CBTKinet, CBUinet, 690, 6160, CBYGinet, CFWHinet	2330	2030
0330	0359	CRI-RTC	1	China Horizons	DL	CRI-RTCinet	2330	1130
0330	0359	CRI-RTC	2	Frontline	GL	CRI-RTCinet	2330	1130
0330	0359	CRI-RTC	3	Biz China	BE	CRI-RTCinet	2330	1130
0330	0359	CRI-RTC	4	In the Spotlight	DL	CRI-RTCinet	2330	1130
0330	0359	CRI-RTC	5	Voices from Other Lands	DL	CRI-RTCinet	2330	1130
0330	0359	CRI-RTC	6	Life in China	DL	CRI-RTCinet	2330	1130
0330	0359	CRI-RTC	7	Listeners' Garden	LI	CRI-RTCinet	2330	1130
0330	0344	DW	1/2	Sports Report	SP	DWinet	2330	0530
0330	0354	PR-EXT	1/4	Network Europe Week	NZ	PR-EXTinet	2330	0530
0330	0354	PR-EXT	2	Network Europe Extra	AC	PR-EXTinet	2330	0530
0330	0339	PR-EXT	3	Around Poland	TR	PR-EXTinet	2330	0530
0330	0339	PR-EXT	5	Letter from Poland	DL	PR-EXTinet	2330	0530
0330	0334	PR-EXT	6	Comment	NC	PR-EXTinet	2330	0530
0330	0349	PR-EXT	7	Europe East	NZ	PR-EXTinet	2330	0530
0330	0359	RNW	1	Curious Orange	DL	RNW2inet	2330	0530
0330	0359	RNW	2/5	Radio Books	LD	RNW2inet	2330	0530
0330	0359	RNW	3	Curious Orange	DL	RNW2inet	2330	0530
0330	0359	RNW	4	The State We're In Midweek Edition	CS	RNW2inet	2330	0530
0330	0359	RNW	6	Earthbeat	EV	RNW2inet	2330	0530
0330	0359	RNW	7	Bridges with Africa	CS	RNW2inet	2330	0530
0330	0359	RNZI	2	New Music Releases	MR	RNZIinet, 15720	2330	1530
0330	0359	RNZI	3	Mailbox (fortnightly)	DX	RNZIinet, 15720	2330	1530
0330	0359	RNZI	3	Spectrum (fortnightly)	DL	RNZIinet, 15720	2330	1530
0330	0359	RNZI	4	Tradewinds	BE	RNZIinet, 15720	2330	1530
0330	0359	RNZI	5	World in Sport	SP	RNZIinet, 15720	2330	1530
0330	0359	RNZI	6	Pacific Correspondent	NA	RNZIinet, 15720	2330	1530
0330	0459	RNZI	7	(see RNZ-NAT)		RNZIinet, 15720	2330	1530
0330	0359	RTE-R1	2	Off the Shelf	AC	RTE-R1inet	2330	0430
0330	0339	VOR-WS	1/4/7	A Stroll Around the Kremlin	AC	6100, 6155, 6240, 7250, 12040, 13735, VOR2inet	2330	0630
0330	0359	VOR-WS	2	Russian by Radio	LL	6100, 6155, 6240, 7250, 12040, 13735, VOR2inet	2330	0630
0330	0359	VOR-WS	3	Folk Box	MF	6100, 6155, 6240, 7250, 12040, 13735, VOR2inet	2330	0630
0330	0359	VOR-WS	5	Moscow Yesterday and Today	AC	6100, 6155, 6240, 7250, 12040, 13735, VOR2inet	2330	0630
0330	0359	VOR-WS	6	Jazz Show	MJ	6100, 6155, 6240, 7250, 12040, 13735, VOR2inet	2330	0630
0330	0359	WRN-NA	1-7	Radio Romania International features	VA	Sirius140, WRN-NAinet	2330	0430
0335	0359	ABC-RN	2/7	The Philosopher's Zone	ED	ABC-RNinet	2335	1335
0335	0354	PR-EXT	6	Multimedia	DX	PR-EXTinet	2335	0535
0340	0404	PR-EXT	3	Talking Jazz	MJ	PR-EXTinet	2340	0540
0340	0359	PR-EXT	5	Studio 15	AC	PR-EXTinet	2340	0540
0340	0344	VOR-WS	1/4/7	Legends of Russian Sports	SP	6100, 6155, 6240, 7250, 12040, 13735, VOR2inet	2340	0640
0345	0359	ABC-RA	7	Ockham's Razor	ST	RAinet	2345	1345
0345	0359	CBC-R1P	4-7	Outfront	DL	CBCVinet, CBTKinet, CBUinet, 690, 6160, CBYGinet, CFWHinet	2345	2045
0345	0359	DW	1	Inspired Minds	GI	DWinet	2345	0545
0345	0359	DW	2	Radio D	LL	DWinet	2345	0545
0345	0359	KBS-WR	2	Faces of Korea	CS	KBS-WR1inet	2345	1245
0345	0359	KBS-WR	3	Business Watch	BE	KBS-WR1inet	2345	1245
0345	0359	KBS-WR	4	Culture on the Move	AC	KBS-WR1inet	2345	1245
0345	0359	KBS-WR	5	Korea Today and Tomorrow	DL	KBS-WR1inet	2345	1245
0345	0359	KBS-WR	6	Seoul Report	DL	KBS-WR1inet	2345	1245
0345	0349	VOR-WS	1	Songs from Russia	MP	6100, 6155, 6240, 7250, 12040, 13735, VOR2inet	2345	0645
0345	0359	VOR-WS	4/7	Musical Tales	LD	6100, 6155, 6240, 7250, 12040, 13735, VOR2inet	2350	0650

UTC Time Start	End	Station/ Network	Day(s)	Program Name	Type	Frequncy/ Platform	EDT	Station Time
0350	0359	PR-EXT	7	Business Week	BE	PR-EXTinet	2350	0550
0350	0359	VOR-WS	1	Russian Hits	MR	6100, 6155, 6240, 7250, 12040, 13735, VOR2inet	2350	0650
0352	0357	CRI-ENG	1-7	Learning Chinese Now	LL	6190, 9690, 9790	2352	1152
0355	0404	PR-EXT	1	A Look at the Weeklies	PR	PR-EXTinet	2355	0555
0355	0404	PR-EXT	2/6	Focus	AC	PR-EXTinet	2355	0555
0355	0404	PR-EXT	4	The Kids	CS	PR-EXTinet	2355	0555
0355	0359	SABC-SAFM	1/7	This New Day	CS	SABC-SAFMinet	2355	0555
0400	0429	ABC-RA	1	In the Loop	CA	RAinet	0000	1400
0400	0429	ABC-RA	7	Australian Bite	DL	RAinet	0000	1400
0400	0459	ABC-RN	1	Hindsight	AC	ABC-RNinet	0000	1400
0400	0419	ABC-RN	2-6	Book Reading	LD	ABC-RNinet	0000	1400
0400	0459	ABC-RN	7	Feature program (from 2/09)	ND	ABC-RNinet	0000	1400
0400	0459	ABC-RN	7	Radio Eye (til 1/09)	ND	ABC-RNinet	0000	1400
0400	0419	BBCWS-AM	1-7	World Briefing	NX	XM131	0000	0500
0400	0419	BBCWS-IE	1-7	World Briefing	NX	BBCWS-IEinet	0000	0500
0400	0419	BBCWS-NX	1-7	World Briefing	NX	BBCWS-NXinet	0000	0500
0400	0419	BBCWS-PR	1-7	World Briefing	NX	Sirius141, NPRfm/am, HD+	0000	0500
0400	0859	CBC-R1A	1/7	CBC Overnight	GI	CBAMinet, CBCTinet, CBDinet, CBHAinet, CBlinet, CBNinet, 640, 6160, CBTinet, CBYinet, CBZFinet, CFGBinet	0000	0100
0400	0844	CBC-R1A	2-6	CBC Overnight	GI	CBAMinet, CBCTinet, CBDinet, CBHAinet, CBlinet, CBNinet, 640, 6160, CBTinet, CBYinet, CBZFinet, CFGBinet	0000	0100
0400	0559	CBC-R1C	2	Tonic	MJ	CBWinet, 990	0000	2300
0400	0459	CBC-R1C	3	Quirks and Quarks	ST	CBWinet, 990	0000	2300
0400	0459	CBC-R1C	4	The Vinyl Cafe	LE	CBWinet, 990	0000	2300
0400	0429	CBC-R1C	5	Afghanada	LD	CBWinet, 990	0000	2300
0400	0459	CBC-R1C	6	Writers and Company	AC	CBWinet, 990	0000	2300
0400	0559	CBC-R1C	7	Vinyl Tap	MP	CBWinet, 990	0000	2300
0400	0459	CBC-R1E	2	Global Arts and Entertainment	AC	CBCLinet, CBCSinet, CBEinet, 1550, CBLAinet, CBMEinet, CBOinet, CBQTinet, CBVEinet, CFFBinet	0000	0000
0400	0429	CBC-R1E	3	From Our Own Correspondent	NA	CBCLinet, CBCSinet, CBEinet, 1550, CBLAinet, CBMEinet, CBOinet, CBQTinet, CBVEinet, CFFBinet	0000	0000
0400	0459	CBC-R1E	4	The Choice	GI	CBCLinet, CBCSinet, CBEinet, 1550, CBLAinet, CBMEinet, CBOinet, CBQTinet, CBVEinet, CFFBinet	0000	0000
0400	0459	CBC-R1E	5	And the Winner Is	FD	CBCLinet, CBCSinet, CBEinet, 1550, CBLAinet, CBMEinet, CBOinet, CBQTinet, CBVEinet, CFFBinet	0000	0000
0400	0459	CBC-R1E	6	Rewind	GI	CBCLinet, CBCSinet, CBEinet, 1550, CBLAinet, CBMEinet, CBOinet, CBQTinet, CBVEinet, CFFBinet	0000	0000
0400	0559	CBC-R1M	1	Saturday Night Blues	MF	CBKinet, 540, CBRinet, 1010, CBXinet, 740, CFYKinet, CHAKinet	0000	2200
0400	0459	CBC-R1M	3-7	Q	AC	CBKinet, 540, CBRinet, 1010, CBXinet, 740, CFYKinet, CHAKinet	0000	2200
0400	0459	CBC-R1P	1	A Propos	MF	CBCVinet, CBTKinet, CBUinet, 690, 6160, CBYGinet, CFWHinet	2300	2100
0400	0559	CBC-R1P	2	In the Key of Charles	MV	CBCVinet, CBTKinet, CBUinet, 690, 6160, CBYGinet, CFWHinet	0000	2100
0400	0459	CBC-R1P	3-7	Ideas	ED	CBCVinet, CBTKinet, CBUinet, 690, 6160, CBYGinet, CFWHinet	0000	2100
0400	0459	CBC-R1S	1	The House	GL	Sirius137	0000	0000
0400	0459	CBC-R1S	3-7	Ideas	ED	Sirius137	0000	0000
0400	0419	CRI-ENG	1/7	News and Reports	NZ	6190	0000	1200
0400	0426	CRI-ENG	2-6	News and Reports	NZ	6190	0000	1200
0400	0429	CRI-RTC	1-7	News and Reports	NZ	CRI-RTCinet	0000	1200
0400	0405	DW	1-7	News	NX	DWinet	0000	0600
0400	0759	ORF-FM4	1-7	Morning Show (in German and English)	MR	ORF-FM4inet	0000	0600
0400	0429	PR-EXT	1	Open Air	CS	PR-EXTinet	0000	0600
0400	0414	PR-EXT	5	Day in the Life	PI	PR-EXTinet	0000	0600
0400	0429	PR-EXT	7	Offside	SP	PR-EXTinet	0000	0600
0400	0404	R.PRG	1/2	News	NX	R.PRGinet, 6080, 6200, 7345	0000	0600
0400	0409	R.PRG	3-7	News and Current Affairs	NX	R.PRGinet, 6080, 6200, 7345	0000	0600
0400	0414	RFI	2-6	African News, Sport, International News	NX	RFImultilingual2inet	0000	0600
0400	0459	RNW	1	The State We're In	CS	RNW2inet, WRNna	0000	0600
0400	0429	RNW	2	Network Europe Extra	AC	RNW2inet, WRNna	0000	0600
0400	0429	RNW	3-7	Newsline	NZ	RNW2inet, WRNna	0000	0600
0400	0759	RNZ-NAT	1	Four Ôtil Eight with Liz Barry	GZ	RNZ-NATinet	0000	1600
0400	0459	RNZ-NAT	7	Music feature	MX	RNZ-NATinet	0000	1600
0400	0659	RNZI	2-6	(see RNZ-NAT)		RNZIinet, 15720, 9615s, 11725w	0000	1600
0400	0444	RTE-R1	1	Conversations with Eamon Dunphy	GI	RTE-R1inet	0000	0500
0400	0429	RTE-R1	2	Farm Week	BE	RTE-R1inet	0000	0500
0400	0429	RTE-R1	6	Seascapes	BE	RTE-R1inet	0000	0500
0400	0429	RTE-R1	7	The Dave Fanning Show	AC	RTE-R1inet	0000	0500
0400	0414	SABC-CHAF	1-7	News	NX	CHAFStudio1inet	0000	0600
0400	0659	SABC-SAFM	1/7	Weekend AM Live	GZ	SABC-SAFMinet	0000	0600
0400	0659	SABC-SAFM	2-6	AM Live	GZ	SABC-SAFMinet	0000	0600
0400	0459	VOR-WS	1	Music and Musicians	MC	6135, 6155, 6240, 7150, 7335, 7350, 9840, 9855, 12030, VOR2inet	0000	0700
0400	0429	VOR-WS	2	This is Russia	DL	6135, 6155, 6240, 7150, 7335, 7350, 9840, 9855, 12030, VOR2inet	0000	0700
0400	0429	VOR-WS	3/6	Encyclopedia "All Russia"	AC	6135, 6155, 6240, 7150, 7335, 7350, 9840, 9855, 12030, VOR2inet	0000	0700
0400	0429	VOR-WS	4/7	Moscow Mailbag	LI	6135, 6155, 6240, 7150, 7335, 7350, 9840, 9855, 12030, VOR2inet	0000	0700
0400	0429	VOR-WS	5	Science Plus	ST	6135, 6155, 6240, 7150, 7335, 7350, 9840, 9855, 12030, VOR2inet	0000	0700
0400	0459	WRN-NA	1-7	Radio Netherlands Worldwide features	VA	Sirius140, WRN-NAinet	0000	0500
0400	0659	WRS-SUI	1	Relay BBC World Service	VA	WRS-SUIinet	0000	0600
0400	0429	WRS-SUI	2-6	BBC World Briefing	NX	WRS-SUIinet	0000	0600
0400	0559	WRS-SUI	7	Relay BBC World Service	NX	WRS-SUIinet	0000	0600
0405	0424	DW	1	In-Box	MB	DWinet	0005	0605

UTC Time Start	End	Station/ Network	Day(s)	Program Name	Type	Frequncy/ Platform	EDT	Station Time
0405	0414	DW	2	Newslink	NZ	DWinet	0005	0605
0405	0429	DW	3-7	Newslink	NZ	DWinet	0005	0605
0405	0429	PR-EXT	2	High Note	MC	PR-EXTinet	0005	0605
0405	0414	PR-EXT	3	The Kids	CS	PR-EXTinet	0005	0605
0405	0429	PR-EXT	4	The Biz	BE	PR-EXTinet	0005	0605
0405	0429	PR-EXT	6	Chart Show	MP	PR-EXTinet	0005	0605
0405	0414	R.PRG	1	Magazine	DL	R.PRGinet, 6080, 6200, 7345	0005	0605
0405	0414	R.PRG	2	Mailbox	LI	R.PRGinet, 6080, 6200, 7345	0005	0605
0405	0459	RNW	1	The State We're In	CS	CBC-R1A	0005	0605
0405	0429	RNW	2	Network Europe Extra	AC	CBC-R1A	0005	0605
0405	0429	RNW	3	Curious Orange	DL	CBC-R1A	0005	0605
0405	0429	RNW	4	The State We're In Midweek Edition	CS	CBC-R1A	0005	0605
0405	0429	RNW	5	Radio Books	LD	CBC-R1A	0005	0605
0405	0429	RNW	6	Earthbeat	EV	CBC-R1A	0005	0605
0405	0429	RNW	7	Bridges with Africa	CS	CBC-R1A	0005	0605
0407	0459	RNZ-NAT	1	The Sunday Feature	GI	RNZ-NATinet	0007	1607
0410	0419	R.PRG	3	One on One	PI	R.PRGinet, 6080, 6200, 7345	0010	0610
0410	0428	R.PRG	4	Talking Point	DL	R.PRGinet, 6080, 6200, 7345	0010	0610
0410	0428	R.PRG	5	Czechs in History (monthly)	AC	R.PRGinet, 6080, 6200, 7345	0010	0610
0410	0428	R.PRG	5	Czechs Today (monthly)	CS	R.PRGinet, 6080, 6200, 7345	0010	0610
0410	0428	R.PRG	5	Spotlight (fortnightly)	TR	R.PRGinet, 6080, 6200, 7345	0010	0610
0410	0418	R.PRG	6	Panorama	CS	R.PRGinet, 6080, 6200, 7345	0010	0610
0410	0418	R.PRG	7	Business News	BE	R.PRGinet, 6080, 6200, 7345	0010	0610
0415	0429	DW	2	Sports Report	SP	DWinet	0015	0615
0415	0429	PR-EXT	3	In Touch	LI	PR-EXTinet	0015	0615
0415	0429	PR-EXT	5	Open Air	CS	PR-EXTinet	0015	0615
0415	0418	R.PRG	1	Sound Czech	LL	R.PRGinet, 6080, 6200, 7345	0015	0615
0415	0419	R.PRG	2	Letter from Prague	DL	R.PRGinet, 6080, 6200, 7345	0015	0615
0415	0419	RFI	2-6	French Press Review	PR	RFImultilingual2inet	0015	0615
0415	0459	SABC-CHAF	1	Straight Talk	NA	CHAFStudio1inet	0015	0615
0415	0439	SABC-CHAF	2/3	Africa Rise and Shine	NZ	CHAFStudio1inet	0015	0615
0415	0459	SABC-CHAF	4-6	Africa Rise and Shine	NZ	CHAFStudio1inet	0015	0615
0415	0459	SABC-CHAF	7	Sport	SP	CHAFStudio1inet	0015	0615
0419	0428	R.PRG	1	One on One	PI	R.PRGinet, 6080, 6200, 7345	0019	0619
0419	0428	R.PRG	2	Czech Books (fortnightly)	AC	R.PRGinet, 6080, 6200, 7345	0019	0619
0419	0428	R.PRG	2	Magic Carpet (monthly)	MZ	R.PRGinet, 6080, 6200, 7345	0019	0619
0419	0428	R.PRG	2	Music Profile (monthly)	MX	R.PRGinet, 6080, 6200, 7345	0019	0619
0419	0428	R.PRG	7	The Arts	AC	R.PRGinet, 6080, 6200, 7345	0019	0619
0420	0459	ABC-RN	2-6	The Daily Planet	MZ	ABC-RNinet	0020	1420
0420	0429	BBC-R4	1-7	Shipping Forecast	WX	BBC-R4inet	0020	0520
0420	0429	BBCWS-AM	1/7	Sports Roundup	SP	XM131	0020	0520
0420	0429	BBCWS-AM	2-6	World Business Report	BE	XM131	0020	0520
0420	0429	BBCWS-IE	1/7	Sports Roundup	SP	BBCWS-IEinet	0020	0520
0420	0429	BBCWS-IE	2-6	World Business Report	BE	BBCWS-IEinet	0020	0520
0420	0429	BBCWS-NX	1/7	Sports Report	SP	BBCWS-NXinet	0020	0520
0420	0429	BBCWS-NX	2-6	World Business Report	BE	BBCWS-NXinet	0020	0520
0420	0429	BBCWS-PR	1	The Instant Guide	NA	Sirius141, NPRfm/am, HD+	0020	0520
0420	0429	BBCWS-PR	2-7	World Business Report	BE	Sirius141, NPRfm/am, HD+	0020	0520
0420	0426	CRI-ENG	1	Reports from Developing Countries	BE	6190	0020	1220
0420	0426	CRI-ENG	7	CRI Roundup	NX	6190	0020	1220
0420	0428	R.PRG	3	Sports News	SP	R.PRGinet, 6080, 6200, 7345	0020	0620
0420	0424	RFI	2-6	Africa Report	NA	RFImultilingual2inet	0020	0620
0425	0429	DW	1	Mission Europe	LL	DWinet	0025	0625
0425	0429	RFI	2-6	Today in France	DL	RFImultilingual2inet	0025	0625
0427	0451	CRI-ENG	1	In the Spotlight	DL	6190	0027	1227
0427	0451	CRI-ENG	2	People in the Know	GL	6190	0027	1227
0427	0451	CRI-ENG	3	Biz China	BE	6190	0027	1227
0427	0451	CRI-ENG	4	China Horizons	DL	6190	0027	1227
0427	0451	CRI-ENG	5	Voices from Other Lands	DL	6190	0027	1227
0427	0451	CRI-ENG	6	Life in China	DL	6190	0027	1227
0427	0451	CRI-ENG	7	Listener's Garden	LI	6190	0027	1227
0430	0459	ABC-RA	1	Talking Point	CA	RAinet	0030	1430
0430	0459	ABC-RA	7	Innovations	ST	RAinet	0030	1430
0430	0442	BBC-R4	1-7	News Briefing	NX	BBC-R4inet	0030	0530
0430	0459	BBCWS-AM	1	Reporting Religion	CS	XM131	0030	0530
0430	0439	BBCWS-AM	2-6	World Briefing	NX	XM131	0030	0530
0430	0459	BBCWS-AM	7	From Our Own Correspondent	NA	XM131	0030	0530
0430	0459	BBCWS-IE	1	Business Weekly	BE	BBCWS-IEinet	0030	0530
0430	0439	BBCWS-IE	2-6	World Briefing	NX	BBCWS-IEinet	0030	0530
0430	0459	BBCWS-IE	7	Politics UK	GL	BBCWS-IEinet	0030	0530
0430	0459	BBCWS-NX	1	Reporting Religion	CS	BBCWS-NXinet	0030	0530
0430	0439	BBCWS-NX	2-6	World Briefing	NX	BBCWS-NXinet	0030	0530
0430	0459	BBCWS-NX	7	From Our Own Correspondent	NA	BBCWS-NXinet	0030	0530
0430	0459	BBCWS-PR	1	From Our Own Correspondent	NA	Sirius141, NPRfm/am, HD+	0030	0530
0430	0439	BBCWS-PR	2-6	World Briefing	NX	Sirius141, NPRfm/am, HD+	0030	0530

UTC Time Start	End	Station/ Network	Day(s)	Program Name	Type	Frequncy/ Platform	EDT	Station Time
0430	0459	BBCWS-PR	7	Global Business	BE	Sirius141, NPRfm/am, HD+	0030	0530
0430	0459	CBC-R1C	5	WireTap	LE	CBWinet, 990	0030	2330
0430	0459	CBC-R1E	3	Culture Shock	CS	CBCLinet, CBCSinet, CBEinet, 1550, CBLAinet, CBMEinet, CBOinet, CBQTinet, CBVEinet, CFFBinet	0030	0030
0430	0459	CRI-RTC	1	China Horizons	DL	CRI-RTCinet	0030	1230
0430	0459	CRI-RTC	2	Frontline	GL	CRI-RTCinet	0030	1230
0430	0459	CRI-RTC	3	Biz China	BE	CRI-RTCinet	0030	1230
0430	0459	CRI-RTC	4	In the Spotlight	DL	CRI-RTCinet	0030	1230
0430	0459	CRI-RTC	5	Voices from Other Lands	DL	CRI-RTCinet	0030	1230
0430	0459	CRI-RTC	6	Life in China	DL	CRI-RTCinet	0030	1230
0430	0459	CRI-RTC	7	Listeners' Garden	LI	CRI-RTCinet	0030	1230
0430	0444	DW	1	Sports Report	SP	DWinet	0030	0630
0430	0459	DW	2	World in Progress	BE	DWinet	0030	0630
0430	0459	DW	3	Spectrum	ST	DWinet	0030	0630
0430	0459	DW	4	Money Talks	BE	DWinet	0030	0630
0430	0459	DW	5	Living Planet	EV	DWinet	0030	0630
0430	0459	DW	6	Inside Europe	NZ	DWinet	0030	0630
0430	0459	DW	7	Insight	NA	DWinet	0030	0630
0430	0404	R.PRG	1/2	News	NX	R.PRGinet	0030	0630
0430	0409	R.PRG	3-7	News and Current Affairs	NX	R.PRGinet	0030	0630
0430	0459	RNW	2	Reloaded	VA	RNW2inet, WRNna, CBC-R1A	0030	0630
0430	0459	RNW	3	Curious Orange	DL	RNW2inet, WRNna	0030	0630
0430	0459	RNW	3-7	Newsline	NZ	CBC-R1A	0030	0630
0430	0459	RNW	4	The State We're In Midweek Edition	CS	RNW2inet, WRNna	0030	0630
0430	0459	RNW	5	Radio Books	LD	RNW2inet, WRNna	0030	0630
0430	0459	RNW	6	Earthbeat	EV	RNW2inet, WRNna	0030	0630
0430	0459	RNW	7	Bridges with Africa	CS	RNW2inet, WRNna	0030	0630
0430	0559	RTE-R1	2-6	Risin' Time	MV	RTE-R1inet	0030	0530
0430	0459	VOR-WS	2	Moscow Yesterday and Today	AC	6135, 6155, 6240, 7150, 7335, 7350, 9840, 9855, 12030, VOR2inet	0030	0730
0430	0449	VOR-WS	3-7	Guest Speaker	PI	6135, 6155, 6240, 7150, 7335, 7350, 9840, 9855, 12030, VOR2inet	0030	0730
0430	0459	VOR-WS	4	Spiritual Flowerbed	CS	6135, 6155, 6240, 7150, 7335, 7350, 9840, 9855, 12030, VOR2inet	0030	0730
0430	0759	WRS-SUI	2-6	Breakfast Show	GZ	WRS-SUInet	0030	0630
0431	0445	WRS-SUI	2-6	Switzerland Today	DL	WRS-SUInet	0031	0631
0435	0414	R.PRG	1	Magazine	DL	R.PRGinet	0035	0635
0435	0414	R.PRG	2	Mailbox	LI	R.PRGinet	0035	0635
0440	0449	BBCWS-AM	2-6	Analysis	NA	XM131	0030	0530
0440	0449	BBCWS-IE	2-6	Analysis	NA	BBCWS-IEinet	0030	0530
0440	0449	BBCWS-NX	2-6	Analysis	NA	BBCWS-NXinet	0030	0530
0440	0459	BBCWS-PR	2-6	Business Daily	BE	Sirius141, NPRfm/am, HD+	0030	0530
0440	0419	R.PRG	3	One on One	PI	R.PRGinet	0040	0640
0440	0428	R.PRG	5	Czechs in History (monthly)	AC	R.PRGinet	0040	0640
0440	0428	R.PRG	5	Czechs Today (monthly)	CS	R.PRGinet	0040	0640
0440	0428	R.PRG	5	Spotlight (fortnightly)	TR	R.PRGinet	0040	0640
0440	0418	R.PRG	6	Panorama	CS	R.PRGinet	0040	0640
0440	0418	R.PRG	7	Business News	BE	R.PRGinet	0040	0640
0440	0459	SABC-CHAF	2	UN Chronicle	NZ	CHAFStudio1inet	0040	0640
0440	0459	SABC-CHAF	3	African Music	MF	CHAFStudio1inet	0040	0640
0443	0444	BBC-R4	1	Bells on Sunday	MX	BBC-R4inet	0043	0543
0443	0444	BBC-R4	2-7	Prayer for the Day	CS	BBC-R4inet	0043	0543
0445	0459	BBC-R4	1	The Other Heartlands	CS	BBC-R4inet	0045	0535
0445	0459	BBC-R4	2-6	Farming Today	BE	BBC-R4inet	0045	0535
0445	0459	BBC-R4	7	iPM	ED	BBC-R4inet	0045	0535
0445	0459	DW	1	Inspired Minds	GI	DWinet	0045	0645
0445	0418	R.PRG	1	Sound Czech	LL	R.PRGinet	0045	0645
0445	0419	R.PRG	2	Letter from Prague	DL	R.PRGinet	0045	0645
0445	0459	RTE-R1	1	The Poem and the Place	LD	RTE-R1inet	0045	0545
0449	0428	R.PRG	1	One on One	PI	R.PRGinet	0049	0649
0449	0428	R.PRG	2	Czech Books (fortnightly)	AC	R.PRGinet	0049	0649
0449	0428	R.PRG	2	Magic Carpet (monthly)	MZ	R.PRGinet	0049	0649
0449	0428	R.PRG	2	Music Profile (monthly)	MX	R.PRGinet	0049	0649
0449	0428	R.PRG	7	The Arts	AC	R.PRGinet	0049	0649
0450	0459	BBCWS-AM	2-6	Sports Roundup	SP	XM131	0030	0530
0450	0459	BBCWS-IE	2-6	Sports Roundup	SP	BBCWS-IEinet	0030	0530
0450	0459	BBCWS-NX	2-6	Sports Roundup	SP	BBCWS-NXinet	0030	0550
0450	0428	R.PRG	3	Sports News	SP	R.PRGinet	0050	0650
0450	0459	VOR-WS	3/5	Spiritual Flowerbed	CS	6135, 6155, 6240, 7150, 7335, 7350, 9840, 9855, 12030, VOR2inet	0050	0750
0450	0459	VOR-WS	6	Russia--1000 Years of Music	MC	6135, 6155, 6240, 7150, 7335, 7350, 9840, 9855, 12030, VOR2inet	0050	0750
0452	0457	CRI-ENG	1-7	Learning Chinese Now	LL	6190	0052	1252
0500	0529	ABC-RA	1	Artworks	AC	RAinet	0100	1500
0500	0529	ABC-RA	2-6	Pacific Beat - Afternoon Edition	NZ	RAinet	0100	1500
0500	0529	ABC-RA	7	Australian Bite	DL	RAinet	0100	1500
0500	0534	ABC-RN	1	Airplay (til 1/09)	LD	ABC-RNinet	0100	1500
0500	0559	ABC-RN	1	Airplay (from 2/09)	LD	ABC-RNinet	0100	1500
0500	0559	ABC-RN	2	AWAYE!	DL	ABC-RNinet	0100	1500
0500	0559	ABC-RN	3	Artworks	AC	ABC-RNinet	0100	1500
0500	0559	ABC-RN	4	By Design	AC	ABC-RNinet	0100	1500

UTC Time Start	End	Station/ Network	Day(s)	Program Name	Type	Frequncy/ Platform	EDT	Station Time
0500	0544	ABC-RN	5/7	PoeticA	LD	ABC-RNinet	0100	1500
0500	0559	ABC-RN	6	Into the Music	MZ	ABC-RNinet	0100	1500
0500	0534	BBC-R4	1	Something Understood	CS	BBC-R4inet	0100	0600
0500	0759	BBC-R4	2-6	Today	NZ	BBC-R4inet	0100	0600
0500	0506	BBC-R4	7	News, Weather, Press Review	NX	BBC-R4inet	0100	0600
0500	0559	BBCWS-AM	1	The Strand	AC	XM131	0100	0600
0500	0559	BBCWS-AM	2	The Forum	CA	XM131	0100	0600
0500	0529	BBCWS-AM	3/5/7	Documentary feature or series	ND	XM131	0100	0600
0500	0529	BBCWS-AM	4	Global Business	BE	XM131	0100	0600
0500	0529	BBCWS-AM	6	Assignment	ND	XM131	0100	0600
0500	0529	BBCWS-IE	1	From Our Own Correspondent	NA	BBCWS-IEinet	0100	0600
0500	0559	BBCWS-IE	2-6	The World Today	NZ	BBCWS-IEinet	0100	0600
0500	0529	BBCWS-IE	7	Assignment	ND	BBCWS-IEinet	0100	0600
0500	0629	BBCWS-NX	1/7	The World Today	NZ	BBCWS-NXinet	0100	0600
0500	0729	BBCWS-NX	2-6	The World Today	NZ	BBCWS-NXinet	0100	0600
0500	0629	BBCWS-PR	1	The World Today	NZ	Sirius141, NPRfm/am, HD+	0100	0600
0500	0729	BBCWS-PR	2-6	The World Today	NZ	Sirius141, NPRfm/am, HD+	0100	0600
0500	0559	BBCWS-PR	7	The World Today	NZ	Sirius141, NPRfm/am, HD+	0100	0600
0500	0559	CBC-R1C	2	Global Arts and Entertainment	AC	CBWinet, 990	0100	0000
0500	0529	CBC-R1C	3	From Our Own Correspondent	NA	CBWinet, 990	0100	0000
0500	0559	CBC-R1C	4	The Choice	GI	CBWinet, 990	0100	0000
0500	0559	CBC-R1C	5	And the Winner Is	FD	CBWinet, 990	0100	0000
0500	0559	CBC-R1C	6	Rewind	GI	CBWinet, 990	0100	0000
0500	0959	CBC-R1E	1/7	CBC Overnight	GI	CBCLinet, CBCSinet, CBEinet, 1550, CBLAinet, CBMEinet, CBOinet, CBQTinet, CBVEinet, CFFBinet	0100	0100
0500	0944	CBC-R1E	2-6	CBC Overnight	GI	CBCLinet, CBCSinet, CBEinet, 1550, CBLAinet, CBMEinet, CBOinet, CBQTinet, CBVEinet, CFFBinet	0100	0100
0500	0659	CBC-R1M	2	Tonic	MJ	CBKinet, 540, CBRinet, 1010, CBXinet, 740, CFYKinet, CHAKinet	0100	2300
0500	0559	CBC-R1M	3	Quirks and Quarks	ST	CBKinet, 540, CBRinet, 1010, CBXinet, 740, CFYKinet, CHAKinet	0100	2300
0500	0559	CBC-R1M	4	The Vinyl Cafe	LE	CBKinet, 540, CBRinet, 1010, CBXinet, 740, CFYKinet, CHAKinet	0100	2300
0500	0529	CBC-R1M	5	Afghanada	LD	CBKinet, 540, CBRinet, 1010, CBXinet, 740, CFYKinet, CHAKinet	0100	2300
0500	0559	CBC-R1M	6	Writers and Company	AC	CBKinet, 540, CBRinet, 1010, CBXinet, 740, CFYKinet, CHAKinet	0100	2300
0500	0659	CBC-R1M	7	Vinyl Tap	MP	CBKinet, 540, CBRinet, 1010, CBXinet, 740, CFYKinet, CHAKinet	0100	2300
0500	0659	CBC-R1P	1	Saturday Night Blues	MF	CBCVinet, CBTKinet, CBUinet, 690, 6160, CBYGinet, CFWHinet	0100	2200
0500	0559	CBC-R1P	3-7	Q	AC	CBCVinet, CBTKinet, CBUinet, 690, 6160, CBYGinet, CFWHinet	0100	2200
0500	0559	CBC-R1S	1	Special Delivery	GI	Sirius137	0100	0100
0500	0559	CBC-R1S	2-7	Rewind	GI	Sirius137	0100	0100
0500	0519	CRI-ENG	1/7	News and Reports	NZ	5960, 6190	0100	1300
0500	0526	CRI-ENG	2-6	News and Reports	NZ	5960, 6190	0100	1300
0500	0529	CRI-RTC	1-7	News and Reports	NZ	CRI-RTCinet	0100	1300
0500	0505	DW	1-7	News	NX	DWinet	0100	0700
0500	0509	KBS-WR	1-7	News	NX	KBS-WR2inet	0100	1400
0500	0509	NHK-RJ	1-7	News	NX	6110	0100	1400
0500	0514	RFI	2-6	African News, Sport, International News	NX	RFImultilingual2inet	0100	0700
0500	0529	RNW	1	The State We're In	CS	RNW2inet	0100	0700
0500	0529	RNW	2	Network Europe Extra	AC	RNW2inet	0100	0700
0500	0529	RNW	3	Earthbeat	EV	RNW2inet	0100	0700
0500	0529	RNW	4	Bridges with Africa	CS	RNW2inet	0100	0700
0500	0529	RNW	5	Curious Orange	DL	RNW2inet	0100	0700
0500	0529	RNW	6	Network Europe Extra	AC	RNW2inet	0100	0700
0500	0529	RNW	7	Radio Books	LD	RNW2inet	0100	0700
0500	0659	RNZ-NAT	2-6	Checkpoint	NZ	RNZ-NATinet	0100	1700
0500	0559	RNZI	7	Tagata o te Moana	DL	RNZIinet, 9615s, 11725w	0100	1700
0500	0654	RTE-R1	1	The Weekend on One	GZ	RTE-R1inet	0100	0600
0500	0559	RTE-R1	7	The Weekend on One	GZ	RTE-R1inet	0100	0600
0500	0514	RTHK-3	1-7	News At One	NX	RTHK-3inet	0100	1300
0500	0514	SABC-CHAF	1-7	News	NX	CHAFStudio1inet	0100	0700
0500	0529	VOR-WS	1/2	Encyclopedia "All Russia"	AC	6135, 7150, 7335, 7350, 9840, 9855, 12030, VOR2inet	0100	0800
0500	0529	VOR-WS	3/6	Moscow Mailbag	LI	6135, 7150, 7335, 7350, 9840, 9855, 12030, VOR2inet	0100	0800
0500	0529	VOR-WS	4	Science Plus	ST	6135, 7150, 7335, 7350, 9840, 9855, 12030, VOR2inet	0100	0800
0500	0559	VOR-WS	5	Music and Musicians	MC	6135, 7150, 7335, 7350, 9840, 9855, 12030, VOR2inet	0100	0800
0500	0529	VOR-WS	7	This is Russia	DL	6135, 7150, 7335, 7350, 9840, 9855, 12030, VOR2inet	0100	0800
0500	0529	WRN-NA	1-7	Israel Radio: News	NX	Sirius140, WRN-NAinet	0100	0600
0505	0514	DW	2	Newslink	NZ	DWinet	0105	0705
0505	0529	DW	3-7	Newslink	NZ	DWinet	0105	0705
0505	0559	RNW	1	The State We're In	CS	CBC-R1E	0105	0705
0505	0529	RNW	2	Network Europe Extra	AC	CBC-R1E	0105	0705
0505	0529	RNW	3	Curious Orange	DL	CBC-R1E	0105	0705
0505	0529	RNW	4	The State We're In Midweek Edition	CS	CBC-R1E	0105	0705
0505	0529	RNW	5	Radio Books	LD	CBC-R1E	0105	0705
0505	0529	RNW	6	Earthbeat	EV	CBC-R1E	0105	0605
0505	0529	RNW	7	Bridges with Africa	CS	CBC-R1E	0105	0705
0507	0534	BBC-R4	7	Open Country	DL	BBC-R4inet	0107	0607
0510	0529	KBS-WR	1	Korean Pop Interactive	MP	KBS-WR2inet	0110	1410
0510	0514	KBS-WR	2-6	News Commentary	NC	KBS-WR2inet	0110	1410
0510	0529	KBS-WR	7	Worldwide Friendship	LI	KBS-WR2inet	0110	1410
0510	0529	NHK-RJ	1	Pop Up Japan	MP	6110	0110	1410

UTC Time Start	End	Station/ Network	Day(s)	Program Name	Type	Frequency/ Platform	EDT	Station Time
0510	0524	NHK-RJ	2/4/6	What's Up Japan	NZ	6110	0110	1410
0510	0529	NHK-RJ	3/5	What's Up Japan	NZ	6110	0110	1410
0510	0529	NHK-RJ	7	World Interactive	LI	6110	0110	1410
0515	0529	DW	2	Sports Report	SP	DWinet	0115	0715
0515	0529	KBS-WR	2	Faces of Korea	CS	KBS-WR2inet	0115	1415
0515	0529	KBS-WR	3	Business Watch	BE	KBS-WR2inet	0115	1415
0515	0529	KBS-WR	4	Culture on the Move	AC	KBS-WR2inet	0115	1415
0515	0529	KBS-WR	5	Korea Today and Tomorrow	DL	KBS-WR2inet	0115	1415
0515	0529	KBS-WR	6	Seoul Report	DL	KBS-WR2inet	0115	1415
0515	0519	RFI	2-6	French Press Review	PR	RFImultilingual2inet	0115	0715
0515	0759	RTHK-3	1	Simon Willson	MP	RTHK-3inet	0115	1315
0515	0659	RTHK-3	2-6	Naked Lunch	GZ	RTHK-3inet	0115	1315
0515	0759	RTHK-3	7	Alyson Hau	MP	RTHK-3inet	0115	1315
0515	0559	SABC-CHAF	1	The Inner Voice	CS	CHAFStudio1inet	0115	0715
0515	0539	SABC-CHAF	2	NEPAD Focus	BE	CHAFStudio1inet	0115	0715
0515	0559	SABC-CHAF	3-6	Africa Rise and Shine	NZ	CHAFStudio1inet	0115	0715
0515	0559	SABC-CHAF	7	37 Degrees	GI	CHAFStudio1inet	0115	0715
0520	0526	CRI-ENG	1	Reports from Developing Countries	BE	5960, 6190	0120	1320
0520	0526	CRI-ENG	7	CRI Roundup	NX	5960, 6190	0120	1320
0520	0524	RFI	2-6	Africa Report	NA	RFImultilingual2inet	0120	0720
0520	0529	WRS-SUI	2-6	Swiss Press Review	PR	WRS-SUInet	0120	0720
0525	0529	NHK-RJ	2/4/6	Easy Japanese	LL	6110	0125	1425
0525	0529	RFI	2-6	Today in France	DL	RFImultilingual2inet	0125	0725
0527	0551	CRI-ENG	1	In the Spotlight	DL	5960, 6190	0127	1327
0527	0551	CRI-ENG	2	People in the Know	GL	5960, 6190	0127	1327
0527	0551	CRI-ENG	3	Biz China	BE	5960, 6190	0127	1327
0527	0551	CRI-ENG	4	China Horizons	DL	5960, 6190	0127	1327
0527	0551	CRI-ENG	5	Voices from Other Lands	DL	5960, 6190	0127	1327
0527	0551	CRI-ENG	6	Life in China	DL	5960, 6190	0127	1327
0527	0551	CRI-ENG	7	Listener's Garden	LI	5960, 6190	0127	1327
0530	0559	ABC-RA	1	Australian Country Style	MW	RAinet	0130	1530
0530	0534	ABC-RA	2/3	Sports Report	SP	RAinet	0130	1530
0530	0559	ABC-RA	4-6	Pacific Beat - On the Mat	CA	RAinet	0130	1530
0530	0559	ABC-RA	7	All in the Mind	HM	RAinet	0130	1530
0530	0559	BBCWS-AM	3-7	The Strand	BE	XM131	0130	0630
0530	0559	BBCWS-IE	1	Heart and Soul	CS	BBCWS-IEinet	0130	0630
0530	0559	BBCWS-IE	7	One Planet	EV	BBCWS-IEinet	0130	0630
0530	0559	CBC-R1C	3	Culture Shock	CS	CBWinet, 990	0130	0030
0530	0559	CBC-R1M	5	WireTap	LE	CBKinet, 540, CBRinet, 1010, CBXinet, 740, CFYKinet, CHAKinet	0130	2330
0530	0559	CRI-RTC	1	China Horizons	DL	CRI-RTCinet	0130	1330
0530	0559	CRI-RTC	2	Frontline	GL	CRI-RTCinet	0130	1330
0530	0559	CRI-RTC	3	Biz China	BE	CRI-RTCinet	0130	1330
0530	0559	CRI-RTC	4	In the Spotlight	DL	CRI-RTCinet	0130	1330
0530	0559	CRI-RTC	5	Voices from Other Lands	DL	CRI-RTCinet	0130	1330
0530	0559	CRI-RTC	6	Life in China	DL	CRI-RTCinet	0130	1330
0530	0559	CRI-RTC	7	Listeners' Garden	LI	CRI-RTCinet	0130	1330
0530	0559	DW	1	Insight	NA	DWinet	0130	0730
0530	0559	DW	2	Eurovox	AC	DWinet	0130	0730
0530	0559	DW	3	Hits in Germany	MP	DWinet	0130	0730
0530	0559	DW	4	Arts on the Air	AC	DWinet	0130	0730
0530	0559	DW	5	Cool	CS	DWinet	0130	0730
0530	0559	DW	6	Dialogue	CS	DWinet	0130	0730
0530	0559	DW	7	A World of Music	MV	DWinet	0130	0730
0530	0559	RNW	2	Reloaded	VA	CBC-R1E	0105	0705
0530	0559	RNW	2/5	Radio Books	LD	RNW2inet	0130	0730
0530	0559	RNW	3	Curious Orange	DL	RNW2inet	0130	0730
0530	0559	RNW	3-7	Newsline	NZ	CBC-R1E	0130	0730
0530	0559	RNW	4	The State We're In Midweek Edition	CS	RNW2inet	0130	0730
0530	0559	RNW	6	Earthbeat	EV	RNW2inet	0130	0730
0530	0559	RNW	7	Bridges with Africa	CS	RNW2inet	0130	0730
0530	0559	RNZ-NAT	7	Tagata o te Moana	DL	RNZ-NATinet	0130	1700
0530	0559	VOR-WS	1	Kaleidoscope	CS	6135, 7150, 7335, 7350, 9840, 9855, 12030, VOR2inet	0130	0830
0530	0559	VOR-WS	2/6	The VOR Treasure Store	AC	6135, 7150, 7335, 7350, 9840, 9855, 12030, VOR2inet	0130	0830
0530	0549	VOR-WS	3	Music Calendar	MX	6135, 7150, 7335, 7350, 9840, 9855, 12030, VOR2inet	0130	0830
0530	0559	VOR-WS	4	Moscow Yesterday and Today	AC	6135, 7150, 7335, 7350, 9840, 9855, 12030, VOR2inet	0130	0830
0530	0559	VOR-WS	7	Timelines	DL	6135, 7150, 7335, 7350, 9840, 9855, 12030, VOR2inet	0130	0830
0530	0529	WRN-NA	1-6	Radio Canada International features	NX	Sirius140, WRN-NAinet	0130	0630
0530	0529	WRN-NA	7	Banns Radio International: Copenhagen Calling	DL	Sirius140, WRN-NAinet	0130	0630
0530	0544	WRS-SUI	2-6	Switzerland Today	DL	WRS-SUInet	0130	0730
0535	0559	ABC-RA	2/3	Pacific Beat - On the Mat	CA	RAinet	0135	1535
0535	0544	ABC-RN	1	Short Story (til 1/09)	LD	ABC-RNinet	0135	1535
0535	0556	BBC-R4	1	On Your Farm	DL	BBC-R4inet	0135	0635
0535	0556	BBC-R4	7	Farming Today This Week	BE	BBC-R4inet	0135	0635
0540	0559	SABC-CHAF	2	UN Chronicle	NZ	CHAFStudio1inet	0140	0740
0545	0559	ABC-RN	1	The Ark (til 1/09)	CS	ABC-RNinet	0145	1545

UTC Time Start	End	Station/ Network	Day(s)	Program Name	Type	Frequncy/ Platform	Station EDT	Time
0545	0559	ABC-RN	5/7	Lingua Franca	AC	ABC-RNinet	0145	1545
0550	0559	VOR-WS	3	Music Around Us	MX	6135, 7150, 7335, 7350, 9840, 9855, 12030, VOR2inet	0150	0850
0550	0559	WRS-SUI	2-6	BBC World Business Report	BE	WRS-SUIinet	0150	0750
0552	0557	CRI-ENG	1-7	Learning Chinese Now	LL	5960, 6190	0152	1352
0557	0559	BBC-R4	1/7	Weather	WX	BBC-R4inet	0157	0657
0600	0629	ABC-RA	1	The Ark (til 1/09)	CS	RAinet	0200	1600
0600	0629	ABC-RA	1	Sunday Profile (from 2/09)	NA	RAinet	0200	1600
0600	0614	ABC-RA	2-6	Sports Report	SP	RAinet	0200	1600
0600	0629	ABC-RA	7	Pacific Review	NZ	RAinet	0200	1600
0600	0659	ABC-RN	1	Music Deli	MZ	ABC-RNinet	0200	1600
0600	0659	ABC-RN	2	Counterpoint	CA	ABC-RNinet	0200	1600
0600	0659	ABC-RN	3-6	Late Night Live	CA	ABC-RNinet	0200	1600
0600	0659	ABC-RN	7	Life and Times	AC	ABC-RNinet	0200	1600
0600	0609	BBC-R4	1	News and Press Review	NX	BBC-R4inet	0200	0700
0600	0759	BBC-R4	7	Today	NZ	BBC-R4inet	0200	0700
0600	0629	BBCWS-AM	2	Heart and Soul	CS	XM131	0200	0700
0600	0629	BBCWS-AM	3-1	The World Today	NZ	XM131	0200	0700
0600	0629	BBCWS-IE	1/7	The World Today	NZ	BBCWS-IEinet	0200	0700
0600	0659	BBCWS-IE	2-6	The World Today	NZ	BBCWS-IEinet	0200	0700
0600	0659	BBCWS-PR	7	The Strand	AC	Sirius141, NPRfm/am, HD+	0200	0700
0600	1059	CBC-R1C	1/7	CBC Overnight	GI	CBWinet, 990	0200	0100
0600	1044	CBC-R1C	2-6	CBC Overnight	GI	CBWinet, 990	0200	0100
0600	0659	CBC-R1M	2	Global Arts and Entertainment	AC	CBKinet, 540, CBRinet, 1010, CBXinet, 740, CFYKinet, CHAKinet	0200	0000
0600	0629	CBC-R1M	3	From Our Own Correspondent	NA	CBKinet, 540, CBRinet, 1010, CBXinet, 740, CFYKinet, CHAKinet	0200	0000
0600	0659	CBC-R1M	4	The Choice	GI	CBKinet, 540, CBRinet, 1010, CBXinet, 740, CFYKinet, CHAKinet	0200	0000
0600	0659	CBC-R1M	5	And the Winner Is	FD	CBKinet, 540, CBRinet, 1010, CBXinet, 740, CFYKinet, CHAKinet	0200	0000
0600	0659	CBC-R1M	6	Rewind	GI	CBKinet, 540, CBRinet, 1010, CBXinet, 740, CFYKinet, CHAKinet	0200	0000
0600	0759	CBC-R1P	2	Tonic	MJ	CBCVinet, CBTKinet, CBUinet, 690, 6160, CBYGinet, CFWHinet	0200	2300
0600	0659	CBC-R1P	3	Quirks and Quarks	ST	CBCVinet, CBTKinet, CBUinet, 690, 6160, CBYGinet, CFWHinet	0200	2300
0600	0659	CBC-R1P	4	The Vinyl Cafe	LE	CBCVinet, CBTKinet, CBUinet, 690, 6160, CBYGinet, CFWHinet	0200	2300
0600	0629	CBC-R1P	5	Afghanada	LD	CBCVinet, CBTKinet, CBUinet, 690, 6160, CBYGinet, CFWHinet	0200	2300
0600	0659	CBC-R1P	6	Writers and Company	AC	CBCVinet, CBTKinet, CBUinet, 690, 6160, CBYGinet, CFWHinet	0200	2300
0600	0759	CBC-R1P	7	Vinyl Tap	MP	CBCVinet, CBTKinet, CBUinet, 690, 6160, CBYGinet, CFWHinet	0200	2300
0600	0659	CBC-R1S	1	Rewind	GI	Sirius137	0200	0200
0600	0659	CBC-R1S	2	Special Delivery	GI	Sirius137	0200	0200
0600	0729	CBC-R1S	3-7	The Point	CA	Sirius137	0200	0200
0600	0619	CRI-ENG	1/7	News and Reports	NZ	6190	0200	1400
0600	0626	CRI-ENG	2-6	News and Reports	NZ	6115	0200	1400
0600	0629	CRI-RTC	1-7	News and Reports	NZ	CRI-RTCinet	0200	1400
0600	0729	CRI-WASH	1-7	China Drive	NZ	CRI-WASHinet	0200	1300
0600	0605	DW	1-7	News	NX	DWinet	0200	0800
0600	0629	RNW	1/3	Earthbeat	EV	RNW2inet	0200	0800
0600	0629	RNW	2	Reloaded	VA	RNW2inet	0200	0800
0600	0629	RNW	4	Bridges with Africa	CS	RNW2inet	0200	0800
0600	0629	RNW	5	Curious Orange	DL	RNW2inet	0200	0800
0600	0629	RNW	6	Network Europe Extra	AC	RNW2inet	0200	0800
0600	0629	RNW	7	Radio Books	LD	RNW2inet	0200	0800
0600	0659	RNZ-NAT	7	Great Encounters	GI	RNZ-NATinet	0200	1800
0600	1259	RNZI	7	(see RNZ-NAT)		RNZIinet, 9615s, 7145s, 9655s	0200	1800
0600	0759	RTE-R1	2-6	Morning Ireland	NZ	RTE-R1inet	0200	0700
0600	0654	RTE-R1	7	Documentary series	GD	RTE-R1inet	0200	0700
0600	0614	SABC-CHAF	1-7	News	NX	CHAFStudio1inet	0200	0800
0600	0629	VOR-WS	1	This is Russia	DL	VOR2inet	0200	0900
0600	0629	VOR-WS	2	Moscow Mailbag	LI	VOR2inet	0200	0900
0600	0629	VOR-WS	3-7	Focus on Asia and the Pacific	NA	VOR2inet	0200	0900
0600	0629	WRN-NA	1-7	Radio Prague features	VA	Sirius140, WRN-NAinet	0200	0700
0600	0659	WRS-SUI	7	Best of the Week	GI	WRS-SUIinet	0200	0800
0605	0624	DW	1	In-Box	MB	DWinet	0205	0805
0605	0614	DW	2	Newslink	NZ	DWinet	0205	0805
0605	0629	DW	3-7	Newslink	NZ	DWinet	0205	0805
0605	0659	RNW	1	The State We're In	CS	CBC-R1C	0205	0805
0605	0629	RNW	2	Network Europe Extra	AC	CBC-R1C	0205	0805
0605	0629	RNW	3	Curious Orange	DL	CBC-R1C	0205	0805
0605	0629	RNW	4	The State We're In Midweek Edition	CS	CBC-R1C	0205	0805
0605	0629	RNW	5	Radio Books	LD	CBC-R1C	0205	0805
0605	0629	RNW	6	Earthbeat	EV	CBC-R1C	0205	0805
0605	0629	RNW	7	Bridges with Africa	CS	CBC-R1C	0205	0805
0610	0654	BBC-R4	1	Sunday	CS	BBC-R4inet	0210	0710
0615	0659	ABC-RA	2/4-6	Talking Point	CA	RAinet	0215	1615
0615	0629	ABC-RA	3	Talking Point	CA	RAinet	0215	1615
0615	0629	DW	2	Sports Report	SP	DWinet	0215	0815
0615	0659	SABC-CHAF	1	Our Heritage	AC	CHAFStudio1inet	0215	0815
0615	0639	SABC-CHAF	2	UN Chronicle	NZ	CHAFStudio1inet	0215	0815
0615	0659	SABC-CHAF	3-6	Africa Rise and Shine	NA	CHAFStudio1inet	0215	0815
0615	0659	SABC-CHAF	7	Tam Tam Express	BE	CHAFStudio1inet	0215	0815

UTC Time Start	End	Station/ Network	Day(s)	Program Name	Type	Frequency/ Platform	EDT	Station Time
0620	0626	CRI-ENG	1	Reports from Developing Countries	BE	6190	0220	1420
0620	0626	CRI-ENG	7	CRI Roundup	NX	6190	0220	1400
0620	0629	WRS-SUI	2-6	World Press Review	PR	WRS-SUInet	0220	0820
0625	0629	DW	1	Mission Europe	LL	DWinet	0225	0825
0627	0651	CRI-ENG	1	In the Spotlight	DL	6190	0227	1427
0627	0651	CRI-ENG	2	People in the Know	GL	6190	0227	1427
0627	0651	CRI-ENG	3	Biz China	BE	6190	0227	1427
0627	0651	CRI-ENG	4	China Horizons	DL	6190	0227	1427
0627	0651	CRI-ENG	5	Voices from Other Lands	DL	6190	0227	1427
0627	0651	CRI-ENG	6	Life in China	DL	6190	0227	1427
0627	0651	CRI-ENG	7	Listener's Garden	LI	6190	0227	1427
0630	0659	ABC-RA	1	MovieTime	AC	RAinet	0230	1630
0630	0659	ABC-RA	3	Law Report	GL	RAinet	0230	1630
0630	0644	ABC-RA	7	Asia Pacific Business	BE	RAinet	0230	1630
0630	0639	BBCWS-AM	1	The Instant Guide	NA	XM131	0230	0730
0630	0659	BBCWS-AM	2	One Planet	EV	XM131	0230	0730
0630	0659	BBCWS-AM	3-7	Outlook	GZ	XM131	0230	0730
0630	0639	BBCWS-IE	1	The Interview	PI	BBCWS-IEinet	0230	0730
0630	0659	BBCWS-IE	7	World Football	SP	BBCWS-IEinet	0230	0730
0630	0659	BBCWS-NX	1	The Interview	PI	BBCWS-NXinet	0230	0730
0630	0659	BBCWS-NX	7	Reporting Religion	CS	BBCWS-NXinet	0230	0730
0630	0659	BBCWS-PR	1	Charlie Gillett's World of Music	MZ	Sirius141, NPRfm/am, HD+	0230	0730
0630	0659	CBC-R1M	1	Culture Shock	CS	CBKinet, 540, CBRinet, 1010, CBXinet, 740, CFYKinet, CHAKinet	0230	0030
0630	0659	CBC-R1P	5	WireTap	LE	CBCVinet, CBTKinet, CBUinet, 690, 6160, CBYGinet, CFWHinet	0230	2330
0630	0654	CRI-RTC	1	China Horizons	DL	CRI-RTCinet	0230	1430
0630	0654	CRI-RTC	2	Frontline	GL	CRI-RTCinet	0230	1430
0630	0654	CRI-RTC	3	Biz China	BE	CRI-RTCinet	0230	1430
0630	0654	CRI-RTC	4	In the Spotlight	DL	CRI-RTCinet	0230	1430
0630	0654	CRI-RTC	5	Voices from Other Lands	DL	CRI-RTCinet	0230	1430
0630	0654	CRI-RTC	6	Life in China	DL	CRI-RTCinet	0230	1430
0630	0654	CRI-RTC	7	Listeners' Garden	LI	CRI-RTCinet	0230	1430
0630	0644	DW	1	Sports Report	SP	DWinet	0230	0830
0630	0659	DW	2	World in Progress	BE	DWinet	0230	0830
0630	0659	DW	3	Spectrum	ST	DWinet	0230	0830
0630	0659	DW	4	Money Talks	BE	DWinet	0230	0830
0630	0659	DW	5	Living Planet	EV	DWinet	0230	0830
0630	0659	DW	6	Inside Europe	NZ	DWinet	0230	0830
0630	0659	DW	7	Insight	NA	DWinet	0230	0830
0630	0639	KBS-WR	1-7	News	NX	KBS-WR2inet	0230	1530
0630	0659	RNW	1/3	Curious Orange	DL	RNW2inet	0230	0830
0630	0659	RNW	2	Reloaded	VA	CBC-R1C	0205	0805
0630	0659	RNW	2/7	Bridges with Africa	VA	RNW2inet	0230	0830
0630	0659	RNW	3-7	Newsline	NZ	CBC-R1C	0230	0830
0630	0659	RNW	4	The State We're In Midweek Edition	CS	RNW2inet	0230	0830
0630	0659	RNW	5	Radio Books	LD	RNW2inet	0230	0830
0630	0659	RNW	6	Earthbeat	EV	RNW2inet	0230	0830
0630	0629	VOR-WS	1	The VOR Treasure Store	AC	VOR2inet	0230	0930
0630	0629	VOR-WS	2/4	Russian by Radio	LL	VOR2inet	0230	0930
0630	0629	VOR-WS	3	Kaleidoscope	CS	VOR2inet	0230	0930
0630	0629	VOR-WS	5	Moscow Yesterday and Today	AC	VOR2inet	0230	0930
0630	0649	VOR-WS	6	Music Calendar	MX	VOR2inet	0230	0930
0630	0659	VOR-WS	7	The Christian Message from Moscow	CS	VOR2inet	0230	0930
0630	0659	WRN-NA	1-7	Radio Sweden features	VA	Sirius140, WRN-NAinet	0230	0730
0630	0644	WRS-SUI	2-6	Switzerland Today	PR	WRS-SUInet	0230	0830
0640	0659	BBCWS-AM	1	Over to You	LI	XM131	0240	0740
0640	0659	KBS-WR	1	Korean Pop Interactive	MP	KBS-WR2inet	0240	1540
0640	0644	KBS-WR	2-6	News Commentary	NC	KBS-WR2inet	0240	1540
0640	0659	KBS-WR	7	Worldwide Friendship	LI	KBS-WR2inet	0240	1540
0640	0659	SABC-CHAF	2	African Music	MF	CHAFStudio1inet	0240	0840
0645	0659	ABC-RA	7	Ockham's Razor	ST	RAinet	0245	1645
0645	0659	DW	1	Inspired Minds	GI	DWinet	0245	0845
0645	0659	KBS-WR	2	Faces of Korea	CS	KBS-WR2inet	0245	1545
0645	0659	KBS-WR	3	Business Watch	BE	KBS-WR2inet	0245	1545
0645	0659	KBS-WR	4	Culture on the Move	AC	KBS-WR2inet	0245	1545
0645	0659	KBS-WR	5	Korea Today and Tomorrow	DL	KBS-WR2inet	0245	1545
0645	0659	KBS-WR	6	Seoul Report	DL	KBS-WR2inet	0245	1545
0650	0659	VOR-WS	6	Music Around Us	MX	VOR2inet	0250	0950
0652	0657	CRI-ENG	1-7	Learning Chinese Now	LL	6190	0252	1452
0655	0657	BBC-R4	1	The Radio 4 Appeal	DL	BBC-R4inet	0255	0755
0655	0659	CRI-RTC	1-7	Chinese Studio	LL	CRI-RTCinet	0230	1430
0655	0659	RTE-R1	1/7	Weather Forecast	WX	RTE-R1inet	0255	0755
0658	0709	BBC-R4	1	Weather, News, Press Review	NX	BBC-R4inet	0258	0758
0700	0729	ABC-RA	2-6	Pacific Beat - Afternoon Edition	NZ	RAinet	0300	1700
0700	0733	ABC-RA	7	Total Rugby	SP	RAinet	0300	1700
0700	0759	ABC-RN	1	Big Ideas	ED	ABC-RNinet	0300	1700

UTC Time Start	End	Station/ Network	Day(s)	Program Name	Type	Frequency/ Platform	Station EDT	Time
0700	0754	ABC-RN	2-6	PM	NZ	ABC-RNinet	0300	1700
0700	0759	ABC-RN	7	Into the Music	MZ	ABC-RNinet	0300	1700
0700	0759	BBCWS-AM	1/7	The World Today	NZ	XM131	0300	0800
0700	0729	BBCWS-AM	2-6	The World Today	NZ	XM131	0300	0800
0700	0759	BBCWS-IE	1/7	The World Today	NZ	BBCWS-IEinet	0300	0800
0700	0729	BBCWS-IE	2-6	The World Today	NZ	BBCWS-IEinet	0300	0800
0700	0759	BBCWS-NX	1/7	The World Today	NZ	BBCWS-NXinet	0300	0800
0700	0759	BBCWS-PR	1/7	The World Today	NZ	Sirius141, NPRfm/am, HD+	0300	0800
0700	1159	CBC-R1M	1/7	CBC Overnight	GI	CBKinet, 540, CBRinet, 1010, CBXinet, 740, CFYKinet, CHAKinet	0300	0100
0700	1144	CBC-R1M	2-6	CBC Overnight	GI	CBKinet, 540, CBRinet, 1010, CBXinet, 740, CFYKinet, CHAKinet	0300	0100
0700	0759	CBC-R1P	2	Global Arts and Entertainment	AC	CBCVinet, CBTKinet, CBUinet, 690, 6160, CBYGinet, CFWHinet	0300	0000
0700	0729	CBC-R1P	3	From Our Own Correspondent	NA	CBCVinet, CBTKinet, CBUinet, 690, 6160, CBYGinet, CFWHinet	0300	0000
0700	0759	CBC-R1P	4	The Choice	GI	CBCVinet, CBTKinet, CBUinet, 690, 6160, CBYGinet, CFWHinet	0300	0000
0700	0759	CBC-R1P	5	And the Winner Is	FD	CBCVinet, CBTKinet, CBUinet, 690, 6160, CBYGinet, CFWHinet	0300	0000
0700	0759	CBC-R1P	6	Rewind	GI	CBCVinet, CBTKinet, CBUinet, 690, 6160, CBYGinet, CFWHinet	0300	0000
0700	0859	CBC-R1S	1	Vinyl Tap	MP	Sirius137	0300	0300
0700	0759	CBC-R1S	2	Deep Roots	MF	Sirius137	0300	0300
0700	0719	CRI-RTC	1/7	News and Reports	NZ	CRI-RTCinet	0300	1500
0700	0754	CRI-RTC	2-6	China Drive	GZ	CRI-RTCinet	0300	1500
0700	0705	DW	1-7	News	NX	DWinet	0300	0900
0700	0704	PR-EXT	1-7	News Bulletin	NX	PR-EXTinet	0300	0900
0700	0714	RFI	2-6	African News, Sport, International News	NX	RFImultilingual2inet	0300	0900
0700	0759	RNW	1	The State We're In	CS	RNW2inet	0300	0900
0700	0729	RNW	2/6	Network Europe Extra	AC	RNW2inet	0300	0900
0700	0729	RNW	3	Earthbeat	EV	RNW2inet	0300	0900
0700	0729	RNW	4	Bridges with Africa	CS	RNW2inet	0300	0900
0700	0729	RNW	5	Curious Orange	DL	RNW2inet	0300	0900
0700	0729	RNW	7	Radio Books	LD	RNW2inet	0300	0900
0700	0959	RNZ-NAT	2-6	Nights with Bryan Crump	VA	RNZ-NATinet	0300	1900
0700	1059	RNZ-NAT	7	Saturday Night with Peter Fry	LE	RNZ-NATinet	0300	1900
0700	0708	RNZI	2-6	Pacific Regional News	NX	RNZIinet, 7145s, 9765w	0300	1900
0700	0709	RTE-R1	1/7	News, Sports, It Says in the Papers	PR	RTE-R1inet	0300	0800
0700	0959	RTHK-3	2-6	Steve James	MP	RTHK-3inet	0300	1500
0700	0714	SABC-CHAF	1-7	News	NX	CHAFStudio1inet	0300	0900
0700	0859	SABC-SAFM	1	Media@SAfm	DX	SABC-SAFMinet	0300	0900
0700	0959	SABC-SAFM	2-6	Morning Talk	NA	SABC-SAFMinet	0300	0900
0700	0959	SABC-SAFM	7	SAfm Lifestyle	DL	SABC-SAFMinet	0300	0900
0700	0729	VOR-WS	1/5	Moscow Mailbag	LI	VOR2inet	0300	1000
0700	0729	VOR-WS	2	Science Plus	ST	VOR2inet	0300	1000
0700	0729	VOR-WS	3/6	This is Russia	DL	VOR2inet	0300	1000
0700	0729	VOR-WS	4/7	Encyclopedia "All Russia"	DAC	VOR2inet	0300	1000
0700	0759	WRN-NA	1-7	Radio Australia features	VA	Sirius140, WRN-NAinet	0300	0800
0700	0759	WRS-SUI	1	Dig It	CS	WRS-SUIinet	0300	0900
0700	0759	WRS-SUI	2/4	Your Space	DL	WRS-SUIinet	0300	0900
0700	0759	WRS-SUI	3	Health Matters	HM	WRS-SUIinet	0300	0900
0700	0759	WRS-SUI	5	Swiss by Design	AC	WRS-SUIinet	0300	0900
0700	0759	WRS-SUI	6	The Classifieds	DL	WRS-SUIinet	0300	0900
0700	0759	WRS-SUI	7	News Week in Review	NX	WRS-SUIinet	0300	0900
0705	0709	ABC-RA	1	Correspondent's Notebook	NA	RAinet	0305	1705
0705	0724	DW	1	In-Box	MB	DWinet	0305	0905
0705	0729	DW	2-6	Newslink	NZ	DWinet	0305	0905
0705	0759	DW	7	Inside Europe	NZ	DWinet	0305	0905
0705	0724	PR-EXT	1	Network Europe Week	NA	PR-EXTinet	0305	0905
0705	0729	PR-EXT	2	Network Europe Week	NZ	PR-EXTinet	0305	0905
0705	0714	PR-EXT	3	Around Poland	TR	PR-EXTinet	0305	0905
0705	0729	PR-EXT	4	Network Europe Extra	AC	PR-EXTinet	0305	0905
0705	0714	PR-EXT	5	Letter from Poland	DL	PR-EXTinet	0305	0905
0705	0724	PR-EXT	6	Multimedia	DX	PR-EXTinet	0305	0905
0705	0724	PR-EXT	7	Europe East	NZ	PR-EXTinet	0305	0905
0705	0759	RNW	1	The State We're In	CS	CBC-R1M	0305	0905
0705	0729	RNW	2	Network Europe Extra	AC	CBC-R1M	0305	0905
0705	0729	RNW	3	Curious Orange	DL	CBC-R1M	0305	0905
0705	0729	RNW	4	The State We're In Midweek Edition	CS	CBC-R1M	0305	0905
0705	0729	RNW	5	Radio Books	LD	CBC-R1M	0305	0905
0705	0729	RNW	6	Earthbeat	EV	CBC-R1M	0305	0905
0705	0729	RNW	7	Bridges with Africa	CS	CBC-R1M	0305	0905
0706	0729	RNZ-NAT	1	One in Five	CS	RNZ-NATinet	0306	1906
0708	0729	RNZI	2-6	Dateline Pacific	NA	RNZIinet, 7145s, 9765s	0308	1908
0710	0724	ABC-RA	1	Ockham's Razor	ST	RAinet	0310	1710
0710	0749	BBC-R4	1	Sunday Worship	CS	BBC-R4inet	0310	0810
0710	0754	RTE-R1	1	Bowman	GI	RTE-R1inet	0310	0810
0710	0759	RTE-R1	7	World Report	NZ	RTE-R1inet	0310	0810
0715	0739	PR-EXT	3	Talking Jazz	MJ	PR-EXTinet	0315	0915
0715	0729	PR-EXT	5	Studio 15	AC	PR-EXTinet	0315	0915
0715	0719	RFI	2-6	French Press Review	PR	RFImultilingual2inet	0315	0915

UTC Time Start	End	Station/ Network	Day(s)	Program Name	Type	Frequency/ Platform	EDT	Station Time
0715	0759	SABC-CHAF	1	Choral Music	MC	CHAFStudio1inet	0315	0915
0715	0759	SABC-CHAF	2	37 Degrees	NZ	CHAFStudio1inet	0315	0915
0715	0759	SABC-CHAF	3	Current Affairs	NA	CHAFStudio1inet	0315	0915
0715	0759	SABC-CHAF	4	Tam Tam Express	GL	CHAFStudio1inet	0315	0915
0715	0759	SABC-CHAF	5	Our Heritage	AC	CHAFStudio1inet	0315	0915
0715	0759	SABC-CHAF	6	Gateway to Africa	BE	CHAFStudio1inet	0315	0915
0715	0759	SABC-CHAF	7	SADC Calling	BE	CHAFStudio1inet	0315	0915
0720	0726	CRI-RTC	1	Reports from Developing Countries	BE	CRI-RTCinet	0320	1520
0720	0726	CRI-RTC	7	CRI Roundup	NX	CRI-RTCinet	0320	1520
0720	0724	RFI	2-6	Africa Report	NA	RFImultilingual2inet	0320	0920
0725	0729	ABC-RA	1	Perspective	NC	RAinet	0325	1725
0725	0729	DW	1	Mission Europe	LL	DWinet	0325	0925
0725	0759	PR-EXT	1	A Look at the Weeklies	PR	PR-EXTinet	0325	0925
0725	0729	PR-EXT	6	Letter from Poland	DL	PR-EXTinet	0325	0925
0725	0759	PR-EXT	7	Offside	SP	PR-EXTinet	0325	0925
0725	0729	RFI	2-6	Today in France	DL	RFImultilingual2inet	0325	0925
0727	0754	CRI-RTC	1/7	China Beat	MP	CRI-RTCinet	0327	1527
0730	0759	ABC-RA	1	Rear Vision	NA	RAinet	0330	1730
0730	0734	ABC-RA	2-6	Sports Report	SP	RAinet	0330	1730
0730	0749	BBCWS-AM	2-6	Business Daily	BE	XM131	0330	0830
0730	0749	BBCWS-IE	2-6	Business Daily	BE	BBCWS-IEinet	0330	0830
0730	0749	BBCWS-NX	2-6	Business Daily	BE	BBCWS-NXinet	0330	0830
0730	0759	BBCWS-PR	2	From Our Own Correspondent	NA	Sirius141, NPRfm/am, HD+	0330	0830
0730	0759	BBCWS-PR	3	Heart and Soul	CS	Sirius141, NPRfm/am, HD+	0330	0830
0730	0759	BBCWS-PR	4	The Interview	PI	Sirius141, NPRfm/am, HD+	0330	0830
0730	0759	BBCWS-PR	5	One Planet	EV	Sirius141, NPRfm/am, HD+	0330	0830
0730	0759	BBCWS-PR	6	Global Business	BE	Sirius141, NPRfm/am, HD+	0330	0830
0730	0759	CBC-R1P	3	Culture Shock	CS	CBCVinet, CBTKinet, CBUinet, 690, 6160, CBYGinet, CFWHinet	0330	0030
0730	0744	CBC-R1S	3-7	Between the Covers	LD	Sirius137	0330	0330
0730	0759	CRI-WASH	1-7	Vogue	CS	CRI-WASHinet	0200	1300
0730	0759	DW	1	A World of Music	MV	DWinet	0330	0930
0730	0759	DW	2	Eurovox	AC	DWinet	0330	0930
0730	0759	DW	3	Hits in Germany	MP	DWinet	0330	0930
0730	0759	DW	4	Arts on the Air	AC	DWinet	0330	0930
0730	0759	DW	5	Cool	CS	DWinet	0330	0930
0730	0759	DW	6	Dialogue	CS	DWinet	0330	0930
0730	0759	PR-EXT	2	High Note	MC	PR-EXTinet	0330	0930
0730	0759	PR-EXT	4	The Biz	BE	PR-EXTinet	0330	0930
0730	0744	PR-EXT	5	Day in the Life	PI	PR-EXTinet	0330	0930
0730	0739	PR-EXT	6	Focus	AC	PR-EXTinet	0330	0930
0730	0759	RFI	2	Network Europe Extra	AC	RFImultilingual2inet	0330	0930
0730	0759	RFI	3	Crossroads	CS	RFImultilingual2inet	0330	0930
0730	0759	RFI	4	Voices	PI	RFImultilingual2inet	0330	0930
0730	0759	RFI	5	Rendezvous	CS	RFImultilingual2inet	0330	0930
0730	0759	RFI	6	World Tracks	MZ	RFImultilingual2inet	0330	0930
0730	0759	RNW	2	Radio Books	LD	RNW2inet	0330	0930
0730	0759	RNW	2	Reloaded	VA	CBC-R1M	0330	0930
0730	0759	RNW	3	Curious Orange	DL	RNW2inet	0330	0930
0730	0759	RNW	3-7	Newsline	NZ	CBC-R1M	0330	0930
0730	0759	RNW	4	The State We're In Midweek Edition	CS	RNW2inet	0330	0930
0730	0759	RNW	5	Radio Books	LD	RNW2inet	0330	0930
0730	0759	RNW	6	Earthbeat	EV	RNW2inet	0330	0930
0730	0759	RNW	7	Bridges with Africa	CS	RNW2inet	0330	0930
0730	0759	RNZ-NAT	1	Te Ahi Kaa	DL	RNZ-NATinet	0330	1930
0730	0759	RNZ-NAT	3	The Sampler	MP	RNZ-NATinet	0330	1930
0730	0759	RNZ-NAT	4	At the Movies	AC	RNZ-NATinet	0330	1930
0730	0759	RNZ-NAT	5	Spectrum	DL	RNZ-NATinet	0330	1930
0730	0759	RNZ-NAT	6	Music feature	MX	RNZ-NATinet	0330	1930
0730	0759	RNZI	2	New Music Releases	MR	RNZIinet, 7145s, 9765w	0330	1930
0730	0759	RNZI	3	Mailbox (fortnightly)	DX	RNZIinet, 7145s, 9765w	0330	1930
0730	0759	RNZI	3	Spectrum (fortnightly)	DL	RNZIinet, 7145s, 9765w	0330	1930
0730	0759	RNZI	4	Tradewinds	BE	RNZIinet, 7145s, 9765w	0330	1930
0730	0759	RNZI	5	World in Sport	SP	RNZIinet, 7145s, 9765w	0330	1930
0730	0759	RNZI	6	Pacific Correspondent	NA	RNZIinet, 7145s, 9765w	0330	1930
0730	0759	VOR-WS	1	Timelines	DL	VOR2inet	0330	1030
0730	0759	VOR-WS	2/6	Kaleidoscope	CS	VOR2inet	0330	1030
0730	0759	VOR-WS	3	Russian by Radio	LL	VOR2inet	0330	1030
0730	0759	VOR-WS	4/5	Jazz Show	MJ	VOR2inet	0330	1030
0730	0759	VOR-WS	7	Folk Box	MF	VOR2inet	0330	1030
0734	0759	ABC-RA	7	Artworks	AC	RAinet	0334	1734
0735	0759	ABC-RA	2-6	Pacific Beat - On the Mat	CA	RAinet	0335	1735
0740	0749	PR-EXT	3	The Kids	CS	PR-EXTinet	0340	0940
0740	0759	PR-EXT	6	Chart Show	MP	PR-EXTinet	0340	0940
0745	0759	CBC-R1S	3-7	Outfront	DL	Sirius137	0345	0345
0745	0759	PR-EXT	5	Open Air	CS	PR-EXTinet	0345	0945

UTC Time Start	End	Station/ Network	Day(s)	Program Name	Type	Frequncy/ Platform	Station EDT	Time
0749	0759	BBCWS-AM	2-6	Analysis	NA	XM131	0349	0849
0749	0759	BBCWS-IE	2-6	Analysis	NA	BBCWS-IEinet	0349	0849
0750	0759	BBC-R4	1	A Point of View	CS	BBC-R4inet	0350	0850
0750	0759	BBCWS-NX	2-6	Analysis	NA	BBCWS-NXinet	0330	0830
0750	0759	PR-EXT	3	In Touch	LI	PR-EXTinet	0350	0950
0755	0759	ABC-RN	2-6	Perspective (til 1/09)	NC	ABC-RNinet	0355	1755
0755	0759	CRI-RTC	1-7	China Studio	LL	CRI-RTCinet	0355	1555
0755	0759	RTE-R1	1	Weather Forecast	WX	RTE-R1inet	0355	0855
0800	0829	ABC-RA	1	Correspondents Report	NA	RAinet	0400	1800
0800	0859	ABC-RA	2-6	PM	NZ	RAinet	0400	1800
0800	0859	ABC-RA	7	The Margaret Throsby Interview	AC	RAinet	0400	1800
0800	0859	ABC-RN	1	The Spirit of Things	CS	ABC-RNinet	0400	1800
0800	0859	ABC-RN	2	Australia Talks (til 1/09)	DL	ABC-RNinet	0400	1800
0800	0859	ABC-RN	2	FORA Radio (from 2/09)	ND	ABC-RNinet	0400	1800
0800	0859	ABC-RN	3-5	Australia Talks	DL	ABC-RNinet	0400	1800
0800	0859	ABC-RN	6	The National Interest	CA	ABC-RNinet	0400	1800
0800	0859	ABC-RN	7	AWAYE!	DL	ABC-RNinet	0400	1800
0800	0859	BBC-R4	1	Broadcasting House	NA	BBC-R4inet	0400	0900
0800	0844	BBC-R4	2	Start the Week	AC	BBC-R4inet	0400	0900
0800	0829	BBC-R4	3	General feature or documentary series	GI	BBC-R4inet	0400	0900
0800	0844	BBC-R4	4	Midweek	CA	BBC-R4inet	0400	0900
0800	0844	BBC-R4	5	In Our Time	ED	BBC-R4inet	0400	0900
0800	0844	BBC-R4	6	Desert Island Discs	LE	BBC-R4inet	0400	0900
0800	0859	BBC-R4	7	Saturday Live	CS	BBC-R4inet	0400	0900
0800	0829	BBCWS-AM	1	Assignment	ND	XM131	0400	0900
0800	0819	BBCWS-AM	2-6	World Briefing	NX	XM131	0400	0900
0800	0829	BBCWS-AM	7	Documentary feature or series	ND	XM131	0400	0900
0800	0859	BBCWS-IE	1	The Forum	CA	BBCWS-IEinet	0400	0900
0800	0819	BBCWS-IE	2-6	World Briefing	NX	BBCWS-IEinet	0400	0900
0800	0859	BBCWS-IE	7	The Strand	AC	BBCWS-IEinet	0400	0900
0800	0829	BBCWS-NX	1	Assignment	ND	BBCWS-NXinet	0400	0900
0800	0819	BBCWS-NX	2-6	World Briefing	NX	BBCWS-NXinet	0400	0900
0800	0829	BBCWS-NX	7	From Our Own Correspondent	NA	BBCWS-NXinet	0400	0900
0800	0829	BBCWS-PR	1	Assignment	ND	Sirius141, NPRfm/am, HD+	0400	0900
0800	0819	BBCWS-PR	2-6	World Briefing	NX	Sirius141, NPRfm/am, HD+	0400	0900
0800	0829	BBCWS-PR	7	From Our Own Correspondent	NA	Sirius141, NPRfm/am, HD+	0400	0900
0800	1259	CBC-R1P	1/7	CBC Overnight	GI	CBCVinet, CBTKinet, CBUinet, 690, 6160, CBYGinet, CFWHinet	0400	0100
0800	1244	CBC-R1P	2-6	CBC Overnight	GI	CBCVinet, CBTKinet, CBUinet, 690, 6160, CBYGinet, CFWHinet	0400	0100
0800	0859	CBC-R1S	2	Inside the Music	MZ	Sirius137	0400	0400
0800	0859	CBC-R1S	3	And The Winner Is	FD	Sirius137	0400	0400
0800	0859	CBC-R1S	4	The Choice	GI	Sirius137	0400	0400
0800	0859	CBC-R1S	5	Dispatches	NA	Sirius137	0400	0400
0800	0859	CBC-R1S	6	Writers and Company	AC	Sirius137	0400	0400
0800	0859	CBC-R1S	7	The Vinyl Cafe	LE	Sirius137	0400	0400
0800	0829	CRI-RTC	1-7	News and Reports	NZ	CRI-RTCinet	0400	1600
0800	1059	CRI-WASH	1-7	Beyond Beijing	MV	CRI-WASHinet	0400	1600
0800	0805	DW	1-7	News	NX	DWinet	0400	1000
0800	0809	KBS-WR	1-7	News	NX	KBS-WR1inet	0400	1700
0800	1059	ORF-FM4	1	Sunny Side Up (in German and English)	MP	ORF-FM4inet	0400	1000
0800	0959	ORF-FM4	2-7	Update (in German and English)	MR	ORF-FM4inet	0400	1000
0800	0804	R.PRG	1/7	News	NX	R.PRGinet	0400	1000
0800	0809	R.PRG	2-6	News and Current Affairs	NX	R.PRGinet	0400	1000
0800	0829	RNW	1/3	Earthbeat	EV	RNW2inet	0400	1000
0800	0829	RNW	2	Reloaded	VA	RNW2inet	0400	1000
0800	0829	RNW	4	Bridges with Africa	CS	RNW2inet	0400	1000
0800	0829	RNW	5	Curious Orange	DL	RNW2inet	0400	1000
0800	0829	RNW	6	Network Europe Extra	AC	RNW2inet	0400	1000
0800	0829	RNW	7	Radio Books	LD	RNW2inet	0400	1000
0800	0959	RNZ-NAT	1	Sounds Historical	AC	RNZ-NATinet	0400	2000
0800	0829	RNZI	2-6	Pacific Music	MF	RNZIinet, 7145s, 9765w	0400	2000
0800	0809	RTE-R1	1/7	News, Sports, It Says in the Papers	PR	RTE-R1inet	0400	0900
0800	0859	RTE-R1	2-6	The Tubridy Show	VA	RTE-R1inet	0400	0900
0800	0959	RTHK-3	1	Sunday PM	GZ	RTHK-3inet	0400	1600
0800	0959	RTHK-3	7	Saturday PM	MP	RTHK-3inet	0400	1600
0800	0829	VOR-WS	1	Science Plus	ST	VOR2inet	0400	1100
0800	0859	VOR-WS	2	Music and Musicians	MC	VOR2inet	0400	1100
0800	0829	VOR-WS	3/5/7	Russia and the World	NA	VOR2inet	0400	1100
0800	0829	VOR-WS	4	This is Russia	DL	VOR2inet	0400	1100
0800	0829	VOR-WS	6	Moscow Mailbag	LI	VOR2inet	0400	1100
0800	0829	WRN-NA	1-7	Voice of Russia features	VA	Sirius140, WRN-NAinet	0400	0900
0800	0829	WRS-SUI	1	Interview of the Week	PI	WRS-SUIinet	0400	1000
0800	1129	WRS-SUI	2-6	On the Beat	GZ	WRS-SUIinet	0400	1000
0800	0859	WRS-SUI	7	Kids in Mind	CS	WRS-SUIinet	0400	1000
0805	0829	DW	2-6	Newslink	NZ	DWinet	0405	1005
0805	0829	DW	7	Network Europe Week	NZ	DWinet	0405	1005

UTC Time Start	End	Station/ Network	Day(s)	Program Name	Type	Frequncy/ Platform	EDT	Station Time
0805	0859	DW	1	Concert Hour	MC	DWinet	0405	1005
0805	0814	R.PRG	1	Mailbox	LI	R.PRGinet	0405	1005
0805	0814	R.PRG	7	Magazine	DL	R.PRGinet	0405	1005
0805	0859	RNW	1	The State We're In	CS	CBC-R1P	0405	1005
0805	0829	RNW	2	Network Europe Extra	AC	CBC-R1P	0405	1005
0805	0829	RNW	3	Curious Orange	DL	CBC-R1P	0405	1005
0805	0829	RNW	4	The State We're In Midweek Edition	CS	CBC-R1P	0405	1005
0805	0829	RNW	5	Radio Books	LD	CBC-R1P	0405	1005
0805	0829	RNW	6	Earthbeat	EV	CBC-R1P	0405	1005
0805	0829	RNW	7	Bridges with Africa	CS	CBC-R1P	0405	1005
0810	0829	KBS-WR	1	Korean Pop Interactive	MP	KBS-WR1inet	0410	1710
0810	0814	KBS-WR	2-6	News Commentary	NC	KBS-WR1inet	0410	1710
0810	0829	KBS-WR	7	Worldwide Friendship	LI	KBS-WR1inet	0410	1710
0810	0819	R.PRG	2	One on One	PI	R.PRGinet	0410	1010
0810	0828	R.PRG	3	Talking Point	DL	R.PRGinet	0410	1010
0810	0828	R.PRG	4	Czechs in History (monthly)	AC	R.PRGinet	0410	1010
0810	0828	R.PRG	4	Czechs Today (monthly)	CS	R.PRGinet	0410	1010
0810	0828	R.PRG	4	Spotlight (fortnightly)	TR	R.PRGinet	0410	1010
0810	0818	R.PRG	5	Panorama	CS	R.PRGinet	0410	1010
0810	0818	R.PRG	6	Business News	BE	R.PRGinet	0410	1010
0810	0859	RTE-R1	1	Sunday Miscellany	LD	RTE-R1inet	0410	0910
0810	0859	RTE-R1	7	Conversations with Eamon Dunphy	PI	RTE-R1inet	0410	0910
0815	0829	KBS-WR	2	Faces of Korea	CS	KBS-WR1inet	0415	1715
0815	0829	KBS-WR	3	Business Watch	BE	KBS-WR1inet	0415	1715
0815	0829	KBS-WR	4	Culture on the Move	AC	KBS-WR1inet	0415	1715
0815	0829	KBS-WR	5	Korea Today and Tomorrow	DL	KBS-WR1inet	0415	1715
0815	0829	KBS-WR	6	Seoul Report	DL	KBS-WR1inet	0415	1715
0815	0819	R.PRG	1	Letter from Prague	DL	R.PRGinet	0415	1015
0815	0818	R.PRG	7	Sound Czech	LL	R.PRGinet	0419	1019
0815	0845	RNZ-NAT	2-5	Windows on the World	ND	RNZ-NATinet	0415	2015
0819	0828	R.PRG	1	Czech Books (fortnightly)	AC	R.PRGinet	0419	1019
0819	0828	R.PRG	1	Magic Carpet (monthly)	MZ	R.PRGinet	0419	1019
0819	0828	R.PRG	1	Music Profile (monthly)	MX	R.PRGinet	0419	1019
0819	0828	R.PRG	6	The Arts	AC	R.PRGinet	0419	1019
0819	0828	R.PRG	7	One on One	PI	R.PRGinet	0419	1019
0820	0829	BBCWS-AM	2-6	World Business Report	BE	XM131	0420	0920
0820	0829	BBCWS-IE	2-6	World Business Report	BE	BBCWS-IEinet	0420	0920
0820	0829	BBCWS-NX	2-6	World Business Report	BE	BBCWS-NXinet	0420	0920
0820	0829	BBCWS-PR	2-6	World Business Report	BE	Sirius141, NPRfm/am, HD+	0420	0920
0820	0828	R.PRG	2	Sports News	SP	R.PRGinet	0420	1020
0830	0859	ABC-RA	1	Innovations	ST	RAinet	0430	1830
0830	0844	BBC-R4	3	Phill Jupitus' Strips	AC	BBC-R4inet	0430	0930
0830	0859	BBCWS-AM	1	Reporting Religion	CS	XM131	0430	0930
0830	0859	BBCWS-AM	2	From Our Own Correspondent	NA	XM131	0430	0930
0830	0859	BBCWS-AM	3	Health Check	HM	XM131	0430	0930
0830	0859	BBCWS-AM	4	Discovery	ST	XM131	0430	0930
0830	0859	BBCWS-AM	5	Digital Planet	DX	XM131	0430	0930
0830	0859	BBCWS-AM	6	One Planet	EV	XM131	0430	0930
0830	0859	BBCWS-AM	7	Science in Action	ST	XM131	0430	0930
0830	0859	BBCWS-IE	2-6	Outlook	GZ	BBCWS-IEinet	0430	0930
0830	0859	BBCWS-NX	1	Reporting Religion	CS	BBCWS-NXinet	0430	0930
0830	0839	BBCWS-NX	2-6	World Briefing	NX	BBCWS-NXinet	0430	0930
0830	0859	BBCWS-NX	7	Global Business	BE	BBCWS-NXinet	0430	0930
0830	0859	BBCWS-PR	1	Reporting Religion	CS	Sirius141, NPRfm/am, HD+	0430	0930
0830	0859	BBCWS-PR	2-6	The Strand	AC	Sirius141, NPRfm/am, HD+	0430	0930
0830	0859	BBCWS-PR	7	Global Business	BE	Sirius141, NPRfm/am, HD+	0430	0930
0830	0859	CRI-RTC	1	China Horizons	DL	CRI-RTCinet	0430	1630
0830	0859	CRI-RTC	2	Frontline	GL	CRI-RTCinet	0430	1630
0830	0859	CRI-RTC	3	Biz China	BE	CRI-RTCinet	0430	1630
0830	0859	CRI-RTC	4	In the Spotlight	DL	CRI-RTCinet	0430	1630
0830	0859	CRI-RTC	5	Voices from Other Lands	DL	CRI-RTCinet	0430	1630
0830	0859	CRI-RTC	6	Life in China	DL	CRI-RTCinet	0430	1630
0830	0859	CRI-RTC	7	Listeners' Garden	LI	CRI-RTCinet	0430	1630
0830	0859	DW	2	World in Progress	BE	DWinet	0430	1030
0830	0859	DW	3	Spectrum	ST	DWinet	0430	1030
0830	0859	DW	4	Money Talks	BE	DWinet	0430	1030
0830	0859	DW	5	Living Planet	EV	DWinet	0430	1030
0830	0859	DW	6	Inside Europe	NZ	DWinet	0430	1030
0830	0859	DW	7	A World of Music	MV	DWinet	0430	1030
0830	0839	KBS-WR	1-7	News	NX	KBS-WR2inet	0430	1730
0830	0859	RNW	1/3	Curious Orange	DL	RNW2inet	0430	1030
0830	0859	RNW	2	Reloaded	VA	CBC-R1P	0430	1030
0830	0859	RNW	2/7	Bridges with Africa	SC	RNW2inet	0430	1030
0830	0859	RNW	3-7	Newsline	NZ	CBC-R1P	0430	1030
0830	0859	RNW	4	The State We're In Midweek Edition	CS	RNW2inet	0430	1030

UTC Time Start	End	Station/ Network	Day(s)	Program Name	Type	Frequency/ Platform	EDT	Station Time
0830	0859	RNW	5	Radio Books	LD	RNW2inet	0430	1030
0830	0859	RNW	6	Earthbeat	EV	RNW2inet	0430	1030
0830	0838	RNZI	2-6	Pacific Regional News	NX	RNZIinet, 7145s, 9765w	0430	2030
0830	0839	VOR-WS	1/4	A Stroll Around the Kremlin	AC	VOR2inet	0430	1130
0830	0859	VOR-WS	3	Folk Box	MF	VOR2inet	0430	1130
0830	0859	VOR-WS	5	The VOR Treasure Store	AC	VOR2inet	0430	1130
0830	0859	VOR-WS	6	Moscow Yesterday and Today	AC	VOR2inet	0430	1130
0830	0859	VOR-WS	7	Kaleidoscope	CS	VOR2inet	0430	1130
0830	0859	WRN-NA	1	Glenn Hauser's World of Radio	DX	Sirius140, WRN-NAinet	0430	0930
0830	0859	WRN-NA	2-6	Radio Canada International features	VA	Sirius140, WRN-NAinet	0430	0930
0830	0844	WRN-NA	7	Radio Guangdong: Guangdong Today	DL	Sirius140, WRN-NAinet	0430	0930
0830	0844	WRS-SUI	3	Stir It Up	CS	WRS-SUIinet	0430	1030
0830	0844	WRS-SUI	4	Gadget Guru	DX	WRS-SUIinet	0430	1030
0830	0844	WRS-SUI	5	Dig It	CS	WRS-SUIinet	0430	1030
0830	0844	WRS-SUI	6	Kids in Mind	CS	WRS-SUIinet	0430	1030
0838	0859	RNZI	2-6	Dateline Pacific	NA	RNZIinet, 7145s, 9765w	0438	2038
0840	0849	BBCWS-NX	2-6	Analysis	NA	BBCWS-NXinet	0440	0940
0840	0929	KBS-WR	1	Korean Pop Interactive	MP	KBS-WR2inet	0440	1740
0840	0844	KBS-WR	2-6	News Commentary	NC	KBS-WR2inet	0440	1740
0840	0929	KBS-WR	7	Worldwide Friendship	LI	KBS-WR2inet	0440	1740
0840	0844	VOR-WS	1/4	Legends of Russian Sports	SP	VOR2inet	0440	1140
0845	0859	BBC-R4	2-6	Book of the Week	LD	BBC-R4inet	0445	0945
0845	1129	CBC-R1A	2-6	Local morning program	NZ	CBAMinet, CBCTinet, CBDinet, CBHAinet, CBIinet, CBNinet, 640, 6160, CBTinet, CBYinet, CBZFinet, CFGBinet	0445	0545
0845	0914	KBS-WR	2-6	Seoul Calling	NZ	KBS-WR2inet	0445	1745
0845	0849	VOR-WS	1	Songs from Russia	MP	VOR2inet	0445	1145
0845	0859	VOR-WS	4	Musical Tales	LD	VOR2inet	0445	1145
0845	0859	WRN-NA	7	UN Radio: Women/Perspective	NZ	Sirius140, WRN-NAinet	0445	0945
0850	0859	BBCWS-NX	2-6	Sports Roundup	SP	BBCWS-NXinet	0450	0950
0850	0859	VOR-WS	1	Russian Hits	MR	VOR2inet	0450	1150
0900	1059	ABC-RA	1	The Music Show	MC	RAinet	0500	1900
0900	0959	ABC-RA	2-6	Australia Talks	DL	RAinet	0500	1900
0900	0929	ABC-RA	7	Asia Review	CA	RAinet	0500	1900
0900	0959	ABC-RN	1	Garrison Keillor's Radio Show	LE	ABC-RNinet	0500	1900
0900	0959	ABC-RN	2	The Science Show	ST	ABC-RNinet	0500	1900
0900	0954	ABC-RN	3	Background Briefing	ND	ABC-RNinet	0500	1900
0900	0959	ABC-RN	4	Encounter	CS	ABC-RNinet	0500	1900
0900	0934	ABC-RN	5	Airplay (from 2/09)	LD	ABC-RNinet	0500	1900
0900	0934	ABC-RN	5	MovieTime (til 1/09)	AC	ABC-RNinet	0500	1900
0900	0934	ABC-RN	6	The Sports Factor (til 1/09)	SP	ABC-RNinet	0500	1900
0900	0934	ABC-RN	6	Verbatim (from 2/09)	AC	ABC-RNinet	0500	1900
0900	0959	ABC-RN	7	Big Ideas (til 1/09)	ED	ABC-RNinet	0500	1900
0900	0959	ABC-RN	7	Saturday Extra (from 2/09)	CA	ABC-RNinet	0500	1900
0900	1014	BBC-R4	1	The Archers Omnibus	LD	BBC-R4inet	0500	1000
0900	0959	BBC-R4	2-6	Woman's Hour	CS	BBC-R4inet	0500	1000
0900	0929	BBC-R4	7	Excess Baggage	TR	BBC-R4inet	0500	1000
0900	0929	BBCWS-AM	1/7	World Briefing	NX	XM131	0500	1000
0900	0959	BBCWS-AM	2-6	World Update	NZ	XM131	0500	1000
0900	0919	BBCWS-IE	1/7	World Briefing	NX	BBCWS-IEinet	0500	1000
0900	0929	BBCWS-IE	2/4/6	Documentary feature or series	ND	BBCWS-IEinet	0500	1000
0900	0929	BBCWS-IE	3	Global Business	BE	BBCWS-IEinet	0500	1000
0900	0929	BBCWS-IE	5	Assignment	ND	BBCWS-IEinet	0500	1000
0900	0929	BBCWS-NX	1/7	World Briefing	NX	BBCWS-NXinet	0500	1000
0900	0959	BBCWS-NX	2-6	World Update	NZ	BBCWS-NXinet	0500	1000
0900	0929	BBCWS-PR	1/7	World Briefing	NX	Sirius141, NPRfm/am, HD+	0500	1000
0900	0959	BBCWS-PR	2-6	World Update	NZ	Sirius141, NPRfm/am, HD+	0500	1000
0900	1159	CBC-R1A	1/7	Local morning program	GZ	CBAMinet, CBCTinet, CBDinet, CBHAinet, CBIinet, CBNinet, 640, 6160, CBTinet, CBYinet, CBZFinet, CFGBinet	0500	0600
0900	0959	CBC-R1S	1	Special Delivery	GI	Sirius137	0500	0500
0900	0959	CBC-R1S	2	Tapestry	CS	Sirius137	0500	0500
0900	0959	CBC-R1S	3-7	Rewind	GI	Sirius137	0500	0500
0900	0929	CRI-RTC	1-7	News and Reports	NZ	CRI-RTCinet	0500	1700
0900	0905	DW	1-7	News	NX	DWinet	0500	1100
0900	0929	RNW	1	Network Europe Extra	AC	RNW2inet	0500	1100
0900	0929	RNW	2	Curious Orange	DL	RNW2inet	0500	1100
0900	0929	RNW	3	The State We're In Midweek Edition	CS	RNW2inet	0500	1100
0900	0929	RNW	4	Radio Books	LD	RNW2inet	0500	1100
0900	0929	RNW	5	Earthbeat	EV	RNW2inet	0500	1100
0900	0929	RNW	6	Bridges with Africa	CS	RNW2inet	0500	1100
0900	0959	RNW	7	The State We're In	CS	RNW2inet	0500	1100
0900	1059	RNZI	2-6	(see RNZ-NAT)		RNZIinet, 7145s, 9765w	0500	2100
0900	0929	RTE-R1	1	The Business	BE	RTE-R1inet	0500	1000
0900	1059	RTE-R1	2-6	Today with Pat Kenny	CA	RTE-R1inet	0500	1000
0900	0959	RTE-R1	7	Playback	GI	RTE-R1inet	0500	1000
0900	0944	RTE-R1X	1	Mass	CS	RTE-R1Xinet	0500	1000
0900	0959	SABC-SAFM	1	Church Service	CS	SABC-SAFMinet	0500	1100
0900	0929	VOR-WS	1/2	This is Russia	DL	VOR2inet	0500	1200

UTC Time Start	End	Station/ Network	Day(s)	Program Name	Type	Frequency/ Platform	EDT	Station Time
0900	0929	VOR-WS	3-7	News and Views	NA	VOR2inet	0500	1200
0900	0929	WRN-NA	1-7	Radio Prague features	VA	Sirius140, WRN-NAinet	0500	1000
0900	0959	WRS-SUI	1	Gadget Guru	DX	WRS-SUIinet	0500	1100
0900	0959	WRS-SUI	7	Swiss By Design	AC	WRS-SUIinet	0500	1100
0904	0930	RNZ-NAT	2	Insight	NA	RNZ-NATinet	0504	2104
0904	0930	RNZ-NAT	3	The Tuesday Feature	GD	RNZ-NATinet	0504	2104
0904	0930	RNZ-NAT	4	The Wednesday Drama	LD	RNZ-NATinet	0504	2104
0904	0930	RNZ-NAT	6	Country Life	DL	RNZ-NATinet	0504	2104
0905	0924	DW	1	In-Box	MB	DWinet	0505	1105
0905	0929	DW	2-6	Newslink	NZ	DWinet	0505	1105
0905	0929	DW	7	Network Europe Week	NZ	DWinet	0505	1105
0906	0930	RNZ-NAT	5	Our Changing World	ST	RNZ-NATinet	0506	2106
0915	0929	KBS-WR	2	Faces of Korea	CS	KBS-WR2inet	0515	1815
0915	0929	KBS-WR	3	Business Watch	BE	KBS-WR2inet	0515	1815
0915	0929	KBS-WR	4	Culture on the Move	AC	KBS-WR2inet	0515	1815
0915	0929	KBS-WR	5	Korea Today and Tomorrow	DL	KBS-WR2inet	0515	1815
0915	0929	KBS-WR	6	Seoul Report	DL	KBS-WR2inet	0515	1815
0920	0929	BBCWS-IE	1/7	Sports Roundup	SP	BBCWS-IEinet	0520	1020
0925	0929	DW	1	Mission Europe	LL	DWinet	0525	1125
0930	0959	ABC-RA	7	Jazz Notes	MJ	RAinet	0530	1930
0930	0959	BBC-R4	7	General feature or documentary series	GI	BBC-R4inet	0500	1000
0930	0959	BBCWS-AM	1	Politics UK	GL	XM131	0530	1030
0930	0959	BBCWS-AM	7	Business Weekly	BE	XM131	0530	1030
0930	0959	BBCWS-IE	1	Politics UK	GL	BBCWS-IEinet	0530	1030
0930	0959	BBCWS-IE	2-6	The Strand	AC	BBCWS-IEinet	0530	1030
0930	0959	BBCWS-IE	7	World Football	SP	BBCWS-IEinet	0530	1030
0930	0959	BBCWS-NX	1	Politics UK	GL	BBCWS-NXinet	0530	1030
0930	0959	BBCWS-NX	7	The Interview	BE	BBCWS-NXinet	0530	1030
0930	0959	BBCWS-PR	1	Business Weekly	BE	Sirius141, NPRfm/am, HD+	0530	1030
0930	0959	BBCWS-PR	7	Politics UK	GL	Sirius141, NPRfm/am, HD+	0530	1030
0930	0954	CRI-RTC	1	China Horizons	DL	CRI-RTCinet	0530	1730
0930	0954	CRI-RTC	2	Frontline	GL	CRI-RTCinet	0530	1730
0930	0954	CRI-RTC	3	Biz China	BE	CRI-RTCinet	0530	1730
0930	0954	CRI-RTC	4	In the Spotlight	DL	CRI-RTCinet	0530	1730
0930	0954	CRI-RTC	5	Voices from Other Lands	DL	CRI-RTCinet	0530	1730
0930	0954	CRI-RTC	6	Life in China	DL	CRI-RTCinet	0530	1730
0930	0954	CRI-RTC	7	Listeners' Garden	LI	CRI-RTCinet	0530	1730
0930	0959	DW	1	World in Progress	BE	DWinet	0530	1130
0930	0959	DW	2	Eurovox	AC	DWinet	0530	1130
0930	0959	DW	3	Hits in Germany	MP	DWinet	0530	1130
0930	0959	DW	4	Arts on the Air	AC	DWinet	0530	1130
0930	0959	DW	5	Cool	CS	DWinet	0530	1130
0930	0959	DW	6	Dialogue	CS	DWinet	0530	1130
0930	0959	DW	7	Insight	NA	DWinet	0530	1130
0930	0959	RNW	1	Reloaded	VA	RNW2inet	0530	1130
0930	0959	RNW	2	Earthbeat	EV	RNW2inet	0530	1130
0930	0959	RNW	3	Bridges with Africa	CS	RNW2inet	0530	1130
0930	0959	RNW	4	Curious Orange	DL	RNW2inet	0530	1130
0930	0959	RNW	5	Network Europe Extra	AC	RNW2inet	0530	1130
0930	0959	RNW	6	Radio Books	LD	RNW2inet	0530	1130
0930	0959	RTE-R1	1	Documentary series	GD	RTE-R1inet	0530	1030
0930	0959	VOR-WS	1	Timelines	DL	VOR2inet	0530	1230
0930	0944	VOR-WS	2	A Stroll Around the Kremlin	AC	VOR2inet	0530	1230
0930	0959	VOR-WS	3	Kaleidoscope	CS	VOR2inet	0530	1230
0930	0959	VOR-WS	4	The VOR Treasure Store	AC	VOR2inet	0530	1230
0930	0959	VOR-WS	5	Folk Box	MF	VOR2inet	0530	1230
0930	0959	VOR-WS	6	Jazz Show	MJ	VOR2inet	0530	1230
0930	0959	VOR-WS	7	The Christian Message from Moscow	CS	VOR2inet	0530	1230
0930	0959	WRN-NA	1-7	KBS World Radio features	VA	Sirius140, WRN-NAinet	0530	1030
0930	0959	WRS-SUI	3	Bookmark (2nd Tues. of the month)	AC	WRS-SUIinet	0530	1130
0935	0959	ABC-RN	5	In Conversation (til 1/09)	ST	ABC-RNinet	0535	1935
0935	0959	ABC-RN	6	The Rhythm Divine (from 2/09)	MF	ABC-RNinet	0535	1935
0945	1229	CBC-R1E	2-6	Local morning program	NZ	CBCLinet, CBCSinet, CBEinet, 1550, CBLAinet, CBMEinet, CBOinet, CBQTinet, CBVEinet, CFFBinet		0545
0945	1029	RTE-R1X	1	Service of the Word	CS	RTE-R1Xinet	0545	0545
0945	0949	VOR-WS	2	Legends of Russian Sports	SP	VOR2inet	0545	1245
0950	0959	VOR-WS	2	Musical Tales	LD	VOR2inet	0550	1250
0955	0959	ABC-RN	3	Perspective (til 1/09)	NC	ABC-RNinet	0555	1955
0955	0959	CRI-RTC	1-7	Chinese Studio	LL	CRI-RTCinet	0530	1730
1000	1029	ABC-RA	2-6	Asia Pacific	NZ	RAinet	0600	2000
1000	1019	ABC-RA	7	Asia Pacific Business	BE	RAinet	0600	2000
1000	1034	ABC-RN	1	Artworks Feature (from 2/09)	AC	ABC-RNinet	0600	2000
1000	1034	ABC-RN	1	Street Stories (til 1/09)	DL	ABC-RNinet	0600	2000
1000	1034	ABC-RN	2	Health Report (til 1/09)	HM	ABC-RNinet	0600	2000
1000	1044	ABC-RN	2-5	The Book Show (from 2/09)	AC	ABC-RNinet	0600	2000
1000	1034	ABC-RN	3	Law Report (til 1/09)	GL	ABC-RNinet	0600	2000

UTC Time Start	End	Station/ Network	Day(s)	Program Name	Type	Frequncy/ Platform	EDT	Station Time
1000	1034	ABC-RN	4	Religion Report (til 1/09)	CS	ABC-RNinet	0600	2000
1000	1034	ABC-RN	5	Media Report (til 1/09)	DX	ABC-RNinet	0600	2000
1000	1059	ABC-RN	6	Music Deli	MZ	ABC-RNinet	0600	2000
1000	1159	ABC-RN	7	The Music Show	MC	ABC-RNinet	0600	2000
1000	1029	BBC-R4	2	General feature or documentary series	GI	BBC-R4inet	0600	1100
1000	1029	BBC-R4	3	World on the Move	ST	BBC-R4inet	0600	1100
1000	1029	BBC-R4	4	History series	AC	BBC-R4inet	0600	1100
1000	1029	BBC-R4	5	From Our Own Correspondent	NA	BBC-R4inet	0600	1100
1000	1029	BBC-R4	6	Documentary feature or series	ND	BBC-R4inet	0600	1100
1000	1029	BBC-R4	7	The Week in Parliament	GL	BBC-R4inet	0600	1100
1000	1019	BBCWS-AM	1/7	World Briefing	NX	XM131	0600	1100
1000	1039	BBCWS-AM	2-6	World Briefing	NX	XM131	0600	1100
1000	1029	BBCWS-IE	1/7	Documentary feature or series	ND	BBCWS-IEinet	0600	1100
1000	1039	BBCWS-IE	2-6	World Briefing	NX	BBCWS-IEinet	0600	1100
1000	1019	BBCWS-NX	1/7	World Briefing	NX	BBCWS-NXinet	0600	1100
1000	1039	BBCWS-NX	2-6	World Briefing	NX	BBCWS-NXinet	0600	1100
1000	1019	BBCWS-PR	1	World Briefing	NX	Sirius141, NPRfm/am, HD+	0600	1100
1000	1029	BBCWS-PR	2-6	World Briefing	NX	Sirius141, NPRfm/am, HD+	0600	1100
1000	1029	BBCWS-PR	7	Global Business	BE	Sirius141, NPRfm/am, HD+	0600	1100
1000	1259	CBC-R1E	1/7	Local morning program	GZ	CBCLinet, CBCSinet, CBEinet, 1550, CBLAinet, CBMEinet, CBOinet, CBQTinet, CBVEinet, CFFBinet	0600	0600
1000	1059	CBC-R1S	1	Inside the Music	MZ	Sirius137	0600	0600
1000	1059	CBC-R1S	2	And The Winner Is	FD	Sirius137	0600	0600
1000	1059	CBC-R1S	3	The Choice	GI	Sirius137	0600	0600
1000	1059	CBC-R1S	4	Dispatches	NA	Sirius137	0600	0600
1000	1059	CBC-R1S	5	Writers and Company	AC	Sirius137	0600	0600
1000	1059	CBC-R1S	6	The Vinyl Cafe	LE	Sirius137	0600	0600
1000	1059	CBC-R1S	7	The Next Chapter	AC	Sirius137	0600	0600
1000	1059	CBC-RCI	1	Masala Canada	GZ	Sirius95	0600	0600
1000	1059	CBC-RCI	2	The Maple Leaf Mailbag	LI	Sirius95	0600	0600
1000	1059	CBC-RCI	3	The Link (second hour)	GZ	Sirius95	0600	0600
1000	1019	CRI-RTC	1/7	News and Reports	NZ	CRI-RTCinet	0600	1800
1000	1054	CRI-RTC	2-6	China Drive	GZ	CRI-RTCinet	0600	1800
1000	1005	DW	1-7	News	NX	DWinet	0600	1200
1000	1159	ORF-FM4	2-7	Reality Check (in German and English)	NZ	ORF-FM4inet	0600	1200
1000	1029	RNW	1	Network Europe Extra	AC	RNW2inet	0600	1200
1000	1029	RNW	2	Curious Orange	DL	RNW2inet	0600	1200
1000	1029	RNW	3	The State We're In Midweek Edition	CS	RNW2inet	0600	1200
1000	1029	RNW	4	Radio Books	LD	RNW2inet	0600	1200
1000	1029	RNW	5	Earthbeat	EV	RNW2inet	0600	1200
1000	1029	RNW	6	Bridges with Africa	CS	RNW2inet	0600	1200
1000	1059	RNW	7	The State We're In	CS	RNW2inet	0600	1200
1000	1039	RNZ-NAT	1	Mediawatch	DX	RNZ-NATinet	0600	2200
1000	1059	RNZ-NAT	2-6	Late Edition	NZ	RNZ-NATinet	0600	2200
1000	1152	RTE-R1	1/7	Marian Finucane	CA	RTE-R1inet	0600	1100
1000	1014	RTHK-3	1	News at Six	NX	RTHK-3inet	0600	1800
1000	1059	RTHK-3	2-6	NewsWrap	NZ	RTHK-3inet	0600	1800
1000	1009	RTHK-3	7	News at Six	NX	RTHK-3inet	0600	1800
1000	1014	SABC-CHAF	1-7	News	NX	CHAFStudio1inet	0600	1200
1000	1059	SABC-SAFM	1	Choral Music	MC	SABC-SAFMinet	0600	1200
1000	1059	SABC-SAFM	2-6	Midday Live	NZ	SABC-SAFMinet	0600	1200
1000	1059	SABC-SAFM	7	Intune	CS	SABC-SAFMinet	0600	1200
1000	1029	WRN-NA	1-7	Radio Romania International features	VA	Sirius140, WRN-NAinet	0600	1100
1000	1859	WRS-SUI	1	Relay BBC World Service	VA	WRS-SUIinet	0600	1200
1000	1029	WRS-SUI	2-6	BBC World Briefing	NX	WRS-SUIinet	0600	1200
1000	1059	WRS-SUI	7	Survival Guide	BE	WRS-SUIinet	0600	1200
1005	1029	DW	2-6	Newslink	NZ	DWinet	0605	1205
1005	1029	DW	7	Network Europe Week	NZ	DWinet	0605	1205
1005	1059	DW		Concert Hour	MC	DWinet	0605	1205
1010	1159	RTHK-3	7	Neil Chase in New York	GZ	RTHK-3inet	0610	1810
1015	1059	BBC-R4	1	Desert Island Discs	LE	BBC-R4inet	0615	1115
1015	1029	RTHK-3	1	Hong Kong Heritage	AC	RTHK-3inet	0615	1815
1015	1059	SABC-CHAF	1	Modern Africa Music	MP	CHAFStudio1inet	0615	1215
1015	1039	SABC-CHAF	2	UN Chronicle	NZ	CHAFStudio1inet	0615	1215
1015	1059	SABC-CHAF	3-6	Current Affairs	NA	CHAFStudio1inet	0615	1215
1015	1059	SABC-CHAF	7	African Music	MP	CHAFStudio1inet	0615	1215
1020	1034	ABC-RA	7	Talking Point	CA	RAinet	0620	2020
1020	1029	BBCWS-AM	1/7	Sports Roundup	SP	XM131	0620	1120
1020	1029	BBCWS-NX	1/7	Sports Roundup	SP	BBCWS-NXinet	0620	1120
1020	1029	BBCWS-PR	1	The Instant Guide	NA	Sirius141, NPRfm/am, HD+	0620	1120
1020	1026	CRI-RTC	1	Reports from Developing Countries	BE	CRI-RTCinet	0620	1820
1020	1026	CRI-RTC	7	CRI Roundup	NX	CRI-RTCinet	0620	1820
1027	1054	CRI-RTC	1/7	China Beat	MP	CRI-RTCinet	0627	1827
1030	1059	ABC-RA	2	Health Report	HM	RAinet	0630	2030
1030	1059	ABC-RA	3	Law Report	GL	RAinet	0630	2030
1030	1059	ABC-RA	4	Rear Vision (from 2/09)	NA	RAinet	0630	2030

UTC Time Start	End	Station/ Network	Day(s)	Program Name	Type	Frequncy/ Platform	EDT	Station Time
1030	1059	ABC-RA	4	Religion Report (til 1/09)	CS	RAinet	0630	2030
1030	1059	ABC-RA	5	Media Report (til 1/09)	DX	RAinet	0630	2030
1030	1059	ABC-RA	5	Futures Report (from 2/09)	ST	RAinet	0630	2030
1030	1059	ABC-RA	6	MovieTime (from 2/09)	AC	RAinet	0630	2030
1030	1059	ABC-RA	6	Sports Factor (til 1/09)	SP	RAinet	0630	2030
1030	1059	BBC-R4	2	Comedy series	LD	BBC-R4inet	0630	1130
1030	1059	BBC-R4	3	Arts or Drama feature or series	AC	BBC-R4inet	0630	1130
1030	1059	BBC-R4	4	Inspector Steine	LD	BBC-R4inet	0630	1130
1030	1059	BBC-R4	5	With Great Pleasure	LD	BBC-R4inet	0630	1130
1030	1059	BBC-R4	6	Arts and drama performance series	LD	BBC-R4inet	0630	1130
1030	1059	BBC-R4	7	From Our Own Correspondent	NA	BBC-R4inet	0630	1130
1030	1059	BBCWS-AM	1	One Planet	EV	XM131	0630	1130
1030	1059	BBCWS-AM	7	Global Business	BE	XM131	0630	1130
1030	1039	BBCWS-IE	1	The Instant Guide	NA	BBCWS-IEinet	0630	1130
1030	1059	BBCWS-IE	7	From Our Own Correspondent	NA	BBCWS-IEinet	0630	1130
1030	1059	BBCWS-NX	1	From Our Own Correspondent	NA	BBCWS-NXinet	0630	1130
1030	1059	BBCWS-NX	7	Reporting Religion	CS	BBCWS-NXinet	0630	1130
1030	1059	BBCWS-PR	1	Science in Action	ST	Sirius141, NPRfm/am, HD+	0630	1130
1030	1059	BBCWS-PR	2-6	The Strand	AC	Sirius141, NPRfm/am, HD+	0630	1130
1030	1059	BBCWS-PR	7	One Planet	EV	Sirius141, NPRfm/am, HD+	0630	1130
1030	1059	DW	2	World in Progress	BE	DWinet	0630	1230
1030	1059	DW	3	Spectrum	ST	DWinet	0630	1230
1030	1059	DW	4	Money Talks	BE	DWinet	0630	1230
1030	1059	DW	5	Living Planet	EV	DWinet	0630	1230
1030	1059	DW	6	Inside Europe	NZ	DWinet	0630	1230
1030	1059	DW	7	A World of Music	MV	DWinet	0630	1230
1030	1129	IRIB	1-7	Koran Reading, News and features	VA	IRIBinet	0630	1400
1030	1059	RNW	1/6	Radio Books	LD	RNW2inet	0630	1230
1030	1059	RNW	2	Earthbeat	EV	RNW2inet	0630	1230
1030	1059	RNW	3	Bridges with Africa	CS	RNW2inet	0630	1230
1030	1059	RNW	4	Curious Orange	DL	RNW2inet	0630	1230
1030	1059	RNW	5	Network Europe Extra	AC	RNW2inet	0630	1230
1030	1153	RTE-R1X	1	Marian Finucane (joined in progress)	CA	RTE-R1Xinet	0630	1130
1030	1044	RTHK-3	1	Reflections from Asia	NA	RTHK-3inet	0630	1830
1030	1044	WRN-NA	1/7	Cmnwlth. B/C Assn: Pick of the Commonwealth	NZ	Sirius140, WRN-NAinet	0630	1130
1030	1059	WRN-NA	2-6	Radio Sweden features	VA	Sirius140, WRN-NAinet	0630	1130
1030	1044	WRS-SUI	2-6	Switzerland Today	DL	WRS-SUIinet	0630	1230
1035	1059	ABC-RA	7	Verbatim	AC	RAinet	0635	2035
1035	1159	ABC-RN	1	The Night Air	LD	ABC-RNinet	0635	2035
1035	1059	ABC-RN	2	Artworks Feature	AC	ABC-RNinet	0635	2035
1035	1059	ABC-RN	3	Rear Vision (til 1/09)	NA	ABC-RNinet	0635	2035
1035	1049	ABC-RN	4	The Ark (til 1/09)	CS	ABC-RNinet	0635	2035
1035	1059	ABC-RN	4	Verbatim (til 1/09)	AC	ABC-RNinet	0635	2035
1040	1049	BBCWS-AM	2-6	Analysis	NA	XM131	0640	1140
1040	1059	BBCWS-IE	1	Over to You	LI	BBCWS-IEinet	0640	1140
1040	1049	BBCWS-IE	2-6	Analysis	NA	BBCWS-IEinet	0640	1140
1040	1049	BBCWS-NX	2-6	Analysis	NA	BBCWS-NXinet	0640	1140
1040	1159	RNZ-NAT	1	Wayne's Music	MP	RNZ-NATinet	0640	2240
1040	1059	SABC-CHAF	2	African Music	MF	CHAFStudio1inet	0640	1240
1045	1059	ABC-RN	2-5	First Person (from 2/09)	LD	ABC-RNinet	0645	2045
1045	1329	CBC-R1C	2-6	Local morning program	NZ	CBWinet, 990	0645	0545
1045	1059	RTHK-3	1	Book Club	LD	RTHK-3inet	0645	1845
1045	1059	WRN-NA	1/7	UN Radio: The UN and Africa	NZ	Sirius140, WRN-NAinet	0645	1145
1045	1059	WRS-SUI	2-6	Cover Story	NA	WRS-SUIinet	0645	1245
1050	1059	ABC-RN	4	Short Story (til 1/09)	LD	ABC-RNinet	0650	2050
1050	1059	BBCWS-AM	2-6	Sports Roundup	SP	XM131	0650	1150
1050	1059	BBCWS-IE	2-6	Sports Roundup	SP	BBCWS-IEinet	0650	1150
1050	1059	BBCWS-NX	2-6	Sports Roundup	SP	BBCWS-NXinet	0650	1150
1055	1059	CRI-RTC	1-7	China Studio	LL	CRI-RTCinet	0655	1855
1100	1129	ABC-RA	1	Sunday Profile	NA	RAinet	0700	2100
1100	1109	ABC-RA	2-6	Sports Report	SP	RAinet	0700	2100
1100	1129	ABC-RA	7	Asia Review	CA	RAinet	0700	2100
1100	1159	ABC-RN	2-5	Life Matters	DL	ABC-RNinet	0700	2100
1100	1134	ABC-RN	6	Airplay (til 1/09)	LD	ABC-RNinet	0700	2100
1100	1159	ABC-RN	6	Airplay (from 2/09)	LD	ABC-RNinet	0700	2100
1100	1129	BBC-R4	1	Quiz or panel game	LE	BBC-R4inet	0700	1200
1100	1156	BBC-R4	2-6	You and Yours	BE	BBC-R4inet	0700	1200
1100	1129	BBC-R4	7	Money Box	BE	BBC-R4inet	0700	1200
1100	1129	BBCWS-AM	1/7	World Briefing	NX	XM131	0700	1200
1100	1119	BBCWS-AM	2-6	World Briefing	NX	XM131	0700	1200
1100	1119	BBCWS-IE	1-7	World Briefing	NX	BBCWS-IEinet	0700	1200
1100	1129	BBCWS-NX	1/7	World Briefing	NX	BBCWS-NXinet	0700	1200
1100	1119	BBCWS-NX	2-6	World Briefing	NX	BBCWS-NXinet	0700	1200
1100	1129	BBCWS-PR	1/7	World Briefing	NX	Sirius141, NPRfm/am, HD+	0700	1200
1100	1119	BBCWS-PR	2-6	World Briefing	NX	Sirius141, NPRfm/am, HD+	0700	1200

UTC Time Start	End	Station/ Network	Day(s)	Program Name	Type	Frequncy/ Platform	Station EDT	Time
1100	1359	CBC-R1C	1/7	Local morning program	GZ	CBWinet, 990	0700	0600
1100	1112	CBC-R1S	1-7	World Report	NX	Sirius137	0700	0700
1100	1119	CRI-ENG	1/7	News and Reports	NZ	5990	0700	1900
1100	1159	CRI-ENG	2-6	China Drive	GZ	5990	0700	1900
1100	1119	CRI-RTC	1/7	News and Reports	NZ	CRI-RTCinet	0700	1900
1100	1154	CRI-RTC	2-6	China Drive	GZ	CRI-RTCinet	0700	1900
1100	1129	CRI-WASH	1-7	Realtime China	NZ	CRI-WASHinet	0700	1900
1100	1105	DW	1-7	News	NX	DWinet	0700	1300
1100	1109	KBS-WR	1-7	News	NX	KBS-WR2inet	0700	2000
1100	1129	RNW	1	Reloaded	VA	RNW2inet	0700	1300
1100	1129	RNW	2	Curious Orange	DL	RNW2inet	0700	1300
1100	1129	RNW	3	The State We're In Midweek Edition	CS	RNW2inet	0700	1300
1100	1129	RNW	4	Radio Books	LD	RNW2inet	0700	1300
1100	1129	RNW	5/7	Earthbeat	EV	RNW2inet	0700	1300
1100	1129	RNW	6	Bridges with Africa	CS	RNW2inet	0700	1300
1100	1159	RNZ-NAT	2	Beale Street Caravan	NJ	RNZ-NATinet	0700	2300
1100	1159	RNZ-NAT	3	Charlie Gillett's World of Music	MZ	RNZ-NATinet	0700	2300
1100	1159	RNZ-NAT	4	Round Midnight with Martin Kwok	MJ	RNZ-NATinet	0700	2300
1100	1159	RNZ-NAT	5	The Music Mix	MR	RNZ-NATinet	0700	2300
1100	1159	RNZ-NAT	6	New Zealand Music Festival or Feature	MZ	RNZ-NATinet	0700	2300
1100	1159	RNZ-NAT	7	Wayne's Music	MP	RNZ-NATinet	0700	2300
1100	1108	RNZI	2-6	Pacific Regional News	NX	RNZIinet, 9655s, 13840w	0700	2300
1100	1159	RTE-R1	2-6	The Ronan Collins Show	MP	RTE-R1inet	0700	1200
1100	1159	RTHK-3	1	Half hour programs in Nepalese and Urdu	GI	RTHK-3inet	0700	1900
1100	1259	RTHK-3	2-6	Peter King	MP	RTHK-3inet	0700	1900
1100	1359	SABC-SAFM	1	SAfm Literature	LD	SABC-SAFMinet	0700	1300
1100	1159	SABC-SAFM	2-6	Otherwise	CS	SABC-SAFMinet	0700	1300
1100	1259	SABC-SAFM	7	African Connection	MF	SABC-SAFMinet	0700	1300
1100	1159	WRN-NA	1-7	Radio Australia features	VA	Sirius140, WRN-NAinet	0700	1200
1100	1659	WRS-SUI	7	Relay BBC World Service	VA	WRS-SUIinet	0700	1300
1105	1124	DW	1	In-Box	MB	DWinet	0705	1305
1105	1129	DW	2-6	Newslink	NZ	DWinet	0705	1305
1105	1159	DW	7	Inside Europe	NZ	DWinet	0705	1305
1108	1129	RNZI	2-6	Dateline Pacific	NA	RNZIinet, 9655s, 13840w	0708	2308
1110	1159	ABC-RA	2-6	PM	NZ	RAinet	0710	2110
1110	1159	KBS-WR	1	Korean Pop Interactive	MP	KBS-WR2inet	0810	2110
1110	1114	KBS-WR	2-6	News Commentary	NC	KBS-WR2inet	0810	2110
1110	1159	KBS-WR	7	Worldwide Friendship	LI	KBS-WR2inet	0810	2110
1113	1359	CBC-R1S	1	The Sunday Edition	CA	Sirius137	0713	0713
1113	1136	CBC-R1S	2-6	The Business Network, Sports and Information	BE	Sirius137	0713	0713
1113	1159	CBC-R1S	7	The House	GL	Sirius137	0713	0713
1115	1144	KBS-WR	2-6	Seoul Calling	NZ	KBS-WR2inet	0815	2115
1120	1129	BBCWS-AM	2-6	World Business Report	BE	XM131	0720	1220
1120	1129	BBCWS-IE	1/7	Sports Roundup	NX	BBCWS-IEinet	0720	1220
1120	1129	BBCWS-IE	2-6	World Business Report	BE	BBCWS-IEinet	0720	1220
1120	1129	BBCWS-NX	2-6	World Business Report	BE	BBCWS-NXinet	0720	1220
1120	1129	BBCWS-PR	2-6	World Business Report	BE	Sirius141, NPRfm/am, HD+	0720	1220
1120	1126	CRI-ENG	1	Reports from Developing Countries	BE	5990	0720	1920
1120	1126	CRI-ENG	7	CRI Roundup	NX	5990	0720	1920
1120	1126	CRI-RTC	1	Reports from Developing Countries	BE	CRI-RTCinet	0720	1920
1120	1126	CRI-RTC	7	CRI Roundup	NX	CRI-RTCinet	0720	1920
1125	1129	DW	1	Mission Europe	LL	DWinet	0725	1325
1127	1151	CRI-ENG	1	China Beat	MP	5990	0727	1927
1127	1151	CRI-ENG	7	Music Memories	MF	5990	0727	1927
1127	1154	CRI-RTC	1/7	China Beat	MP	CRI-RTCinet	0727	1927
1130	1159	ABC-RA	1	Speaking Out	DL	RAinet	0730	2130
1130	1159	ABC-RA	7	All in the Mind	HM	RAinet	0730	2130
1130	1156	BBC-R4	1	The Food Programme	CS	BBC-R4inet	0730	1230
1130	1156	BBC-R4	7	The News Quiz	LE	BBC-R4inet	0730	1230
1130	1159	BBCWS-AM	1	Reporting Religion	CS	XM131	0730	1230
1130	1139	BBCWS-AM	2-6	World Briefing	NX	XM131	0730	1230
1130	1159	BBCWS-AM	7	World Football	SP	XM131	0730	1230
1130	1159	BBCWS-IE	1	Reporting Religion	CS	BBCWS-IEinet	0730	1230
1130	1139	BBCWS-IE	2-6	World Briefing	NX	BBCWS-IEinet	0730	1230
1130	1159	BBCWS-IE	7	The Interview	PI	BBCWS-IEinet	0730	1230
1130	1159	BBCWS-NX	1	The Interview	PI	BBCWS-NXinet	0730	1230
1130	1139	BBCWS-NX	2-6	World Briefing	NX	BBCWS-NXinet	0730	1230
1130	1159	BBCWS-NX	7	Politics UK	GL	BBCWS-NXinet	0730	1230
1130	1159	BBCWS-PR	1	From Our Own Correspondent	NA	Sirius141, NPRfm/am, HD+	0730	1230
1130	1139	BBCWS-PR	2-6	World Briefing	NX	Sirius141, NPRfm/am, HD+	0730	1230
1130	1159	BBCWS-PR	7	The Interview	PI	Sirius141, NPRfm/am, HD+	0730	1230
1130	1259	CBC-R1A	2-6	The Current	NA	CBAMinet, CBCTinet, CBDinet, CBHAinet, CBIinet, CBNinet, 640, 6160, CBTinet, CBYinet, CBZFinet, CFGBinet	0730	0830
1130	1159	CRI-WASH	1-7	News and Reports	NZ	CRI-WASHinet	0730	1930
1130	1159	DW	1	A World of Music	MV	DWinet	0730	1330
1130	1159	DW	2	Eurovox	AC	DWinet	0730	1330

UTC Time Start	End	Station/ Network	Day(s)	Program Name	Type	Frequncy/ Platform	EDT	Station Time
1130	1159	DW	3	Hits in Germany	MP	DWinet	0730	1330
1130	1159	DW	4	Arts on the Air	AC	DWinet	0730	1330
1130	1159	DW	5	Cool	CS	DWinet	0730	1330
1130	1159	DW	6	Dialogue	CS	DWinet	0730	1330
1130	1229	IRIB	1-7	Koran Reading, News and features	VA	IRIBinet	0730	1500
1130	1134	R.PRG	1/7	News	NX	R.PRGinet	0730	1330
1130	1139	R.PRG	2-6	News and Current Affairs	NX	R.PRGinet	0730	1330
1130	1159	RNW	1/3	Bridges with Africa	CS	RNW2inet	0730	1330
1130	1159	RNW	2	Earthbeat	EV	RNW2inet	0730	1330
1130	1159	RNW	4/7	Curious Orange	DL	RNW2inet	0730	1330
1130	1159	RNW	5	Network Europe Extra	AC	RNW2inet	0730	1330
1130	1159	RNW	6	Radio Books	LD	RNW2inet	0730	1330
1130	1159	RNZI	2	Mailbox (fortnightly)	DX	RNZIinet, 9655s, 13840w	0730	2330
1130	1159	RNZI	2	Spectrum (fortnightly)	DL	RNZIinet, 9655s, 13840w	0730	2330
1130	1159	RNZI	3	Tradewinds	BE	RNZIinet, 9655s, 13840w	0730	2330
1130	1159	RNZI	4	World in Sport	SP	RNZIinet, 9655s, 13840w	0730	2330
1130	1159	RNZI	5	Pacific Correspondent	NA	RNZIinet, 9655s, 13840w	0730	2330
1130	1159	RNZI	6	Waiata	MF	RNZIinet, 9655s, 13840w	0730	2330
1130	1144	WRS-SUI	2-6	Switzerland Today	DL	WRS-SUIinet	0730	1330
1135	1159	ABC-RN	6	The Night Air (til 1/09)	LD	ABC-RNinet	0735	2135
1135	1144	R.PRG	1	Mailbox	LI	R.PRGinet	0735	1335
1135	1144	R.PRG	7	Magazine	DL	R.PRGinet	0735	1335
1137	1259	CBC-R1S	2-6	The Current	NA	Sirius137	0737	0737
1140	1159	BBCWS-AM	2-6	Business Daily	BE	XM131	0740	1240
1140	1159	BBCWS-IE	2-6	Business Daily	BE	BBCWS-IEinet	0740	1240
1140	1149	BBCWS-NX	2-6	Analysis	NA	BBCWS-NXinet	0740	1240
1140	1159	BBCWS-PR	2-6	Business Daily	BE	Sirius141, NPRfm/am, HD+	0740	1240
1140	1149	R.PRG	2	One on One	PI	R.PRGinet	0740	1340
1140	1158	R.PRG	3	Talking Point	DL	R.PRGinet	0740	1340
1140	1158	R.PRG	4	Czechs in History (monthly)	AC	R.PRGinet	0740	1340
1140	1158	R.PRG	4	Czechs Today (monthly)	CS	R.PRGinet	0740	1340
1140	1158	R.PRG	4	Spotlight (fortnightly)	TR	R.PRGinet	0740	1340
1140	1148	R.PRG	5	Panorama	CS	R.PRGinet	0740	1340
1140	1148	R.PRG	6	Business News	BE	R.PRGinet	0740	1340
1145	1429	CBC-R1M	2-6	Local morning program	NZ	CBKinet, 540, CBRinet, 1010, CBXinet, 740, CFYKinet, CHAKinet	0745	0545
1145	1159	KBS-WR	2	Faces of Korea	CS	KBS-WR2inet	0845	2145
1145	1159	KBS-WR	3	Business Watch	BE	KBS-WR2inet	0845	2145
1145	1159	KBS-WR	4	Culture on the Move	AC	KBS-WR2inet	0845	2145
1145	1159	KBS-WR	5	Korea Today and Tomorrow	DL	KBS-WR2inet	0845	2145
1145	1159	KBS-WR	6	Seoul Report	DL	KBS-WR2inet	0845	2145
1145	1149	R.PRG	1	Letter from Prague	DL	R.PRGinet	0745	1345
1145	1148	R.PRG	7	Sound Czech	LL	R.PRGinet	0749	1349
1149	1158	R.PRG	1	Czech Books (fortnightly)	AC	R.PRGinet	0749	1349
1149	1158	R.PRG	1	Magic Carpet (monthly)	MZ	R.PRGinet	0749	1349
1149	1158	R.PRG	1	Music Profile (monthly)	MX	R.PRGinet	0749	1349
1149	1158	R.PRG	6	The Arts	AC	R.PRGinet	0749	1349
1149	1158	R.PRG	7	One on One	PI	R.PRGinet	0749	1349
1150	1159	BBCWS-NX	2-6	Sports Roundup	SP	BBCWS-NXinet	0750	1250
1150	1158	R.PRG	2	Sports News	SP	R.PRGinet	0750	1350
1152	1157	CRI-ENG	1/7	Learning Chinese Now	LL	5990	0727	1927
1153	1159	RTE-R1	1/7	Weather and Sea Area Forecast	WX	RTE-R1inet	0753	1253
1155	1159	CRI-RTC	1-7	China Studio	LL	CRI-RTCinet	0755	1955
1157	1159	BBC-R4	1-7	Weather	WX	BBC-R4inet	0757	1257
1200	1559	ABC-RA	1	Sunday Night Talk	CS	RAinet	0800	2200
1200	1259	ABC-RA	2-6	Late Night Live	CA	RAinet	0800	2200
1200	1559	ABC-RA	7	Saturday Night Country	MW	RAinet	0800	2200
1200	1359	ABC-RN	1/7	The Weekend Planet	MZ	ABC-RNinet	0800	2200
1200	1259	ABC-RN	2-6	Late Night Live	CA	ABC-RNinet	0800	2200
1200	1229	BBC-R4	1	The World This Weekend	NX	BBC-R4inet	0800	1300
1200	1229	BBC-R4	2-6	The World at One	NX	BBC-R4inet	0800	1300
1200	1209	BBC-R4	7	News	NX	BBC-R4inet	0800	1300
1200	1259	BBCWS-AM	1-7	Newshour	NZ	XM131	0800	1300
1200	1259	BBCWS-IE	1/7	Newshour	NZ	BBCWS-IEinet	0800	1300
1200	1229	BBCWS-IE	2/4/6	Documentary feature or series	ND	BBCWS-IEinet	0800	1300
1200	1229	BBCWS-IE	3	Global Business	BE	BBCWS-IEinet	0800	1300
1200	1229	BBCWS-IE	5	Assignment	ND	BBCWS-IEinet	0800	1300
1200	1259	BBCWS-NX	1/7	Newshour	NZ	BBCWS-NXinet	0800	1300
1200	1359	BBCWS-NX	2-6	Newshour	NZ	BBCWS-NXinet	0800	1300
1200	1259	BBCWS-PR	1/7	Newshour	NZ	Sirius141, NPRfm/am, HD+	0800	1300
1200	1359	BBCWS-PR	2-6	Newshour	NZ	Sirius141, NPRfm/am, HD+	0800	1300
1200	1459	CBC-R1A	1	The Sunday Edition	CA	CBAMinet, CBCTinet, CBDinet, CBHAinet, CBIinet, CBNinet, 640, 6160, CBTinet, CBYinet, CBZFinet, CFGBinet	0800	0900
1200	1259	CBC-R1A	7	The House	GL	CBAMinet, CBCTinet, CBDinet, CBHAinet, CBIinet, CBNinet, 640, 6160, CBTinet, CBYinet, CBZFinet, CFGBinet	0800	0900
1200	1459	CBC-R1M	1/7	Local morning program	GZ	CBKinet, 540, CBRinet, 1010, CBXinet, 740, CFYKinet, CHAKinet	0800	0600
1200	1259	CBC-R1S	7	Quirks and Quarks	ST	Sirius137	0800	0800
1200	1229	CRI-RTC	1-7	News and Reports	NZ	CRI-RTCinet	0800	2000

UTC Time Start	End	Station/ Network	Day(s)	Program Name	Type	Frequncy/ Platform	EDT	Station Time
1200	1214	CRI-WASH	1-7	China Biz Report	BE	CRI-WASHinet	0800	2000
1200	1205	DW	1-7	News	NX	DWinet	0800	1400
1200	1209	KBS-WR	1-7	News	NX	KBS-WR2inet, 9650	0800	2100
1200	1209	NHK-RJ	1-7	News	NX	6120	0800	2100
1200	1204	PR-EXT	1-7	News Bulletin	NX	PR-EXTinet	0800	1400
1200	1229	RNW	1	Network Europe Extra	AC	RNW2inet, WRNna	0800	1400
1200	1229	RNW	2	Curious Orange	DL	RNW2inet, WRNna	0800	1400
1200	1229	RNW	3	The State We're In Midweek Edition	CS	RNW2inet, WRNna	0800	1400
1200	1229	RNW	4	Radio Books	LD	RNW2inet, WRNna	0800	1400
1200	1229	RNW	5	Earthbeat	EV	RNW2inet, WRNna	0800	1400
1200	1229	RNW	6	Bridges with Africa	CS	RNW2inet, WRNna	0800	1400
1200	1259	RNW	7	The State We're In	CS	RNW2inet, WRNna	0800	1400
1200	1759	RNZ-NAT	1-7	All Night Programme	VA	RNZ-NATinet	0800	0000
1200	1259	RNZI	2-6	(see RNZ-NAT)		RNZIinet, 9655s, 13840w	0800	0000
1200	1259	RTE-R1	1	This Week	NA	RTE-R1inet	0800	1300
1200	1244	RTE-R1	2-6	News At One	NZ	RTE-R1inet	0800	1300
1200	1259	RTE-R1	7	Saturdayview	GL	RTE-R1inet	0800	1300
1200	1359	RTHK-3	1	Pete's Quiet Night In	MP	RTHK-3inet	0800	2000
1200	1359	RTHK-3	7	World Vibes	MZ	RTHK-3inet	0800	2000
1200	1359	SABC-SAFM	2-6	Afternoon Talk	NA	SABC-SAFMinet	0800	1400
1200	1259	WRN-NA	1-7	Radio Netherlands features	VA	Sirius140, WRN-NAinet	0800	1300
1200	1359	WRS-SUI	2-6	Relay BBC World Service	VA	WRS-SUIinet	0800	1400
1204	1229	RNZ-NAT	1-7	Music from Midnight	MP	RNZ-NATinet	0800	0000
1205	1229	DW	2-6	Newslink	NZ	DWinet	0805	1405
1205	1229	DW	7	Network Europe Week	NZ	DWinet	0805	1405
1205	1259	DW	1	Concert Hour	MC	DWinet	0805	1405
1205	1229	PR-EXT	1	Europe East	NZ	PR-EXTinet	0805	1405
1205	1229	PR-EXT	2	Focus	AC	PR-EXTinet	0805	1405
1205	1229	PR-EXT	3	Day in the Life	PI	PR-EXTinet	0805	1405
1205	1229	PR-EXT	4	Around Poland	TR	PR-EXTinet	0805	1405
1205	1229	PR-EXT	5	Focus	AC	PR-EXTinet	0805	1405
1205	1229	PR-EXT	6	Business Week	BE	PR-EXTinet	0805	1405
1205	1224	PR-EXT	7	Network Europe Week	NA	PR-EXTinet	0805	1405
1210	1259	BBC-R4	7	Any Questions?	CA	BBC-R4inet	0810	1310
1210	1259	KBS-WR	1	Korean Pop Interactive	MP	KBS-WR2inet	0810	2110
1210	1259	KBS-WR	1	Korean Pop Interactive	MP	9650	0840	2140
1210	1214	KBS-WR	2-6	News Commentary	NC	KBS-WR2inet, 9650	0810	2110
1210	1214	KBS-WR	2-6	News Commentary	NC	KBS-WR2inet, 9650	0840	2140
1210	1229	KBS-WR	7	Worldwide Friendship	LI	KBS-WR2inet	0810	2110
1210	1259	KBS-WR	7	Worldwide Friendship	LI	9650	0840	2140
1210	1229	NHK-RJ	1	Pop Up Japan	MP	6120	0810	2110
1210	1224	NHK-RJ	2/4/6	What's Up Japan	NZ	6120	0810	2110
1210	1229	NHK-RJ	3/5	What's Up Japan	NZ	6120	0810	2110
1210	1229	NHK-RJ	7	World Interactive	LI	6120	0810	2110
1215	1229	CRI-WASH	1-7	People in the Know	GL	CRI-WASHinet	0815	2015
1215	1229	KBS-WR	2	Faces of Korea	CS	KBS-WR2inet	0815	2115
1215	1244	KBS-WR	2-6	Seoul Calling	NZ	9650	0815	2115
1215	1229	KBS-WR	3	Business Watch	BE	KBS-WR2inet	0815	2115
1215	1229	KBS-WR	4	Culture on the Move	AC	KBS-WR2inet	0815	2115
1215	1229	KBS-WR	5	Korea Today and Tomorrow	DL	KBS-WR2inet	0815	2115
1215	1229	KBS-WR	6	Seoul Report	DL	KBS-WR2inet	0815	2115
1225	1229	NHK-RJ	2/4/6	Easy Japanese	LL	6120	0825	2125
1225	1239	PR-EXT	7	A Look at the Weeklies	PR	PR-EXTinet	0825	1425
1230	1259	BBC-R4	1	Documentary feature or series	ND	BBC-R4inet	0830	1330
1230	1259	BBC-R4	2	Quiz or panel game	LE	BBC-R4inet	0830	1330
1230	1259	BBC-R4	3	Soul Music	MD	BBC-R4inet	0830	1330
1230	1259	BBC-R4	4	The Media Show	DX	BBC-R4inet	0830	1330
1230	1259	BBC-R4	5	Open Country	DL	BBC-R4inet	0830	1330
1230	1259	BBC-R4	6	Feedback	LI	BBC-R4inet	0830	1330
1230	1259	BBCWS-IE	2	Health Check	HM	BBCWS-IEinet	0830	1330
1230	1259	BBCWS-IE	3	Digital Planet	DX	BBCWS-IEinet	0830	1330
1230	1259	BBCWS-IE	4	Discovery	ST	BBCWS-IEinet	0830	1330
1230	1259	BBCWS-IE	5	One Planet	EV	BBCWS-IEinet	0830	1330
1230	1259	BBCWS-IE	6	Science in Action	ST	BBCWS-IEinet	0830	1330
1230	1359	CBC-R1E	2-6	The Current	NA	CBCLinet, CBCSinet, CBEinet, 1550, CBLAinet, CBMEinet, CBOinet, CBQTinet, CBVEinet, CFFBinet	0830	0830
1230	1259	CRI-RTC	1	China Horizons	DL	CRI-RTCinet	0830	2030
1230	1259	CRI-RTC	2	Frontline	GL	CRI-RTCinet	0830	2030
1230	1259	CRI-RTC	3	Biz China	BE	CRI-RTCinet	0830	2030
1230	1259	CRI-RTC	4	In the Spotlight	DL	CRI-RTCinet	0830	2030
1230	1259	CRI-RTC	5	Voices from Other Lands	DL	CRI-RTCinet	0830	2030
1230	1259	CRI-RTC	6	Life in China	DL	CRI-RTCinet	0830	2030
1230	1259	CRI-RTC	7	Listeners' Garden	LI	CRI-RTCinet	0830	2030
1230	1259	CRI-WASH	1-7	News and Reports	NZ	CRI-WASHinet	0830	2030
1230	1259	DW	2	World in Progress	BE	DWinet	0830	1430
1230	1259	DW	3	Spectrum	ST	DWinet	0830	1430

UTC Time Start	End	Station/ Network	Day(s)	Program Name	Type	Frequncy/ Platform	EDT	Station Time
1230	1259	DW	4	Money Talks	BE	DWinet	0830	1430
1230	1259	DW	5	Living Planet	EV	DWinet	0830	1430
1230	1259	DW	6	Inside Europe	NZ	DWinet	0830	1430
1230	1259	DW	7	Insight	NA	DWinet	0830	1430
1230	1259	PR-EXT	1	In Touch	LI	PR-EXTinet	0830	1430
1230	1259	PR-EXT	2/5	Talking Jazz	MJ	PR-EXTinet	0830	1430
1230	1259	PR-EXT	3	The Biz	BE	PR-EXTinet	0830	1430
1230	1259	PR-EXT	4	Studio 15	AC	PR-EXTinet	0830	1430
1230	1259	PR-EXT	6	Offside	SP	PR-EXTinet	0830	1430
1230	1259	RNW	1	Reloaded	VA	RNW2inet, WRNna	0830	1430
1230	1259	RNW	2	Earthbeat	EV	RNW2inet, WRNna	0830	1430
1230	1259	RNW	3	Bridges with Africa	CS	RNW2inet, WRNna	0830	1430
1230	1259	RNW	4	Curious Orange	DL	RNW2inet, WRNna	0830	1430
1230	1259	RNW	5	Network Europe Extra	AC	RNW2inet, WRNna	0830	1430
1230	1259	RNW	6	Radio Books	LD	RNW2inet, WRNna	0830	1430
1230	1259	RNZ-NAT	1	Discovery	ST	RNZ-NATinet	0830	0030
1230	1259	RNZ-NAT	2	At the Movies	AC	RNZ-NATinet	0830	0030
1230	1259	RNZ-NAT	3	Insight	NA	RNZ-NATinet	0830	0030
1230	1259	RNZ-NAT	4	Spectrum	DL	RNZ-NATinet	0830	0030
1230	1259	RNZ-NAT	5	One in Five	CS	RNZ-NATinet	0830	0030
1230	1259	RNZ-NAT	6	Laugh Track	LE	RNZ-NATinet	0830	0030
1230	1259	RNZ-NAT	7	History Repeated	AC	RNZ-NATinet	0830	0030
1240	1259	PR-EXT	7	Chart Show	MP	PR-EXTinet	0840	1440
1245	1529	CBC-R1P	2-6	Local morning program	NZ	CBCVinet, CBTKinet, CBUinet, 690, 6160, CBYGinet, CFWHinet	0845	0545
1245	1259	KBS-WR	2	Faces of Korea	CS	9650	0845	2145
1245	1259	KBS-WR	3	Business Watch	BE	9650	0845	2145
1245	1259	KBS-WR	4	Culture on the Move	AC	9650	0845	2145
1245	1259	KBS-WR	5	Korea Today and Tomorrow	DL	9650	0845	2145
1245	1259	KBS-WR	6	Seoul Report	DL	9650	0845	2145
1245	1359	RTE-R1	2-6	Liveline	LI	RTE-R1inet	0845	1345
1300	1329	ABC-RA	2-6	Asia Pacific	NZ	RAinet	0900	2300
1300	1319	ABC-RN	2-6	Book Reading	LD	ABC-RNinet	0900	2300
1300	1344	BBC-R4	1	Gardeners' Question Time	DL	BBC-R4inet	0900	1400
1300	1314	BBC-R4	2-6	The Archers	LD	BBC-R4inet	0900	1400
1300	1329	BBC-R4	7	Any Answers?	CA	BBC-R4inet	0900	1400
1300	1329	BBCWS-AM	1	From Our Own Correspondent	NA	XM131	0900	1400
1300	1329	BBCWS-AM	2-6	Outlook	GZ	XM131	0900	1400
1300	1359	BBCWS-AM	7	The Strand	AC	XM131	0900	1400
1300	1359	BBCWS-IE	1	The Forum	CA	BBCWS-IEinet	0900	1400
1300	1359	BBCWS-IE	2-6	Newshour	NZ	BBCWS-IEinet	0900	1400
1300	1329	BBCWS-NX	1	Global Business	BE	BBCWS-NXinet	0900	1400
1300	1329	BBCWS-NX	7	Assignment	ND	BBCWS-NXinet	0900	1400
1300	1329	BBCWS-PR	1	Global Business	BE	Sirius141, NPRfm/am, HD+	0900	1400
1300	1329	BBCWS-PR	7	Assignment	ND	Sirius141, NPRfm/am, HD+	0900	1400
1300	1359	CBC-R1A	2-5	Q	AC	CBAMinet, CBCTinet, CBDinet, CBHAinet, CBIinet, CBNinet, 640, 6160, CBTinet, CBYinet, CBZFinet, CFGBinet	0900	1000
1300	1429	CBC-R1A	6	Q	AC	CBAMinet, CBCTinet, CBDinet, CBHAinet, CBIinet, CBNinet, 640, 6160, CBTinet, CBYinet, CBZFinet, CFGBinet	0900	1000
1300	1429	CBC-R1A	7	Go!	LE	CBAMinet, CBCTinet, CBDinet, CBHAinet, CBIinet, CBNinet, 640, 6160, CBTinet, CBYinet, CBZFinet, CFGBinet	0900	1000
1300	1559	CBC-R1E	1	The Sunday Edition	CA	CBCLinet, CBCSinet, CBEinet, 1550, CBLAinet, CBMEinet, CBOinet, CBQTinet, CBVEinet, CFFBinet	0900	0900
1300	1359	CBC-R1E	7	The House	GL	CBCLinet, CBCSinet, CBEinet, 1550, CBLAinet, CBMEinet, CBOinet, CBQTinet, CBVEinet, CFFBinet	0900	0900
1300	1559	CBC-R1P	1/7	Local morning program	GZ	CBCVinet, CBTKinet, CBUinet, 690, 6160, CBYGinet, CFWHinet	0900	0600
1300	1436	CBC-R1S	2-6	Q	AC	Sirius137	0900	0900
1300	1429	CBC-R1S	7	Go!	LE	Sirius137	0900	0900
1300	1326	CRI-ENG	1-5	News and Reports	NZ	9570, 11885, 15230	0900	2100
1300	1319	CRI-ENG	6/7	News and Reports	NZ	9570, 11885, 15230	0900	2100
1300	1329	CRI-RTC	1-7	News and Reports	NZ	CRI-RTCinet	0900	2100
1300	1329	CRI-WASH	1/6	Life in China	DL	CRI-WASHinet	0900	2100
1300	1329	CRI-WASH	2	Frontline	GL	CRI-WASHinet	0900	2100
1300	1329	CRI-WASH	3	Biz China	BE	CRI-WASHinet	0900	2100
1300	1329	CRI-WASH	4	In the Spotlight	DL	CRI-WASHinet	0900	2100
1300	1329	CRI-WASH	5	Voices from Other Lands	CS	CRI-WASHinet	0900	2100
1300	1329	CRI-WASH	7	China Horizons	DL	CRI-WASHinet	0900	2100
1300	1305	DW	1-7	News	NX	DWinet	0900	1500
1300	1305	RNW	1/7	News	NX	RNW2inet	0900	1500
1300	1329	RNW	2-6	Newsline	NZ	RNW2inet	0900	1500
1300	1359	RNZI	1/7	Tagata o te Moana	DL	RNZIinet, 6170s, 13840w	0900	0100
1300	1308	RNZI	2-6	Pacific Regional News	NX	RNZIinet, 6170s, 13840w	0900	0100
1300	1652	RTE-R1	1	Sunday Sport with Adrian Eames	SP	RTE-R1inet	0900	1400
1300	1659	RTE-R1	7	Saturday Sport with John Kenny	GL	RTE-R1inet	0900	1400
1300	1359	RTHK-3	2-6	Teen Time	CS	RTHK-3inet	0900	2100
1300	1659	SABC-SAFM	7	SAfm Sports Special	SP	SABC-SAFMinet	0900	1500

UTC Time Start	End	Station/ Network	Day(s)	Program Name	Type	Frequncy/ Platform	EDT	Station Time
1300	1359	WRN-NA	1	RTE Ireland: Sunday Miscellany	LI	Sirius140, WRN-NAinet	0900	1400
1300	1359	WRN-NA	2-6	RTE Ireland: The Tubridy Show	CA	Sirius140, WRN-NAinet	0900	1400
1300	1359	WRN-NA	7	RTE Ireland: Conversations with Eamon Dunphy	PI	Sirius140, WRN-NAinet	0900	1400
1305	1324	DW	1	In-Box	LI	DWinet	0905	1505
1305	1329	DW	2-6	Newslink	NZ	DWinet	0905	1505
1305	1359	DW	7	Inside Europe	NZ	DWinet	0905	1505
1305	1329	RNW	1	Network Europe Extra	AC	RNW2inet	0905	1505
1305	1359	RNW	7	Network Europe Week	NA	RNW2inet	0905	1505
1305	1329	RNZ-NAT	4	BBC Feature	GI	RNZ-NATinet	0905	0105
1305	1329	RNZ-NAT	5	Ideas	ED	RNZ-NATinet	0905	0105
1308	1329	RNZI	2-6	Dateline Pacific	NA	RNZIinet, 6170s, 13840w	0908	0108
1315	1359	BBC-R4	2-6	The Afternoon Play	LD	BBC-R4inet	0915	1415
1315	1344	RNZ-NAT	1	Te Ahi Kaa	DL	RNZ-NATinet	0915	0115
1315	1344	RNZ-NAT	2	One Planet	EV	RNZ-NATinet	0915	0115
1315	1344	RNZ-NAT	3	BBC Feature	GI	RNZ-NATinet	0915	0115
1315	1359	RNZ-NAT	6	From the World	GI	RNZ-NATinet	0915	0115
1315	1344	RNZ-NAT	7	Our Changing World	ST	RNZ-NATinet	0915	0115
1320	1359	ABC-RN	2-5	The Daily Planet	MZ	ABC-RNinet	0920	2320
1320	1459	ABC-RN	6	Sound Quality	MZ	ABC-RNinet	0920	2320
1320	1326	CRI-ENG	6	CRI Roundup	NX	9570, 11885, 15230	0920	2120
1320	1326	CRI-ENG	7	Reports from Developing Countries	BE	9570, 11885, 15230	0920	2120
1325	1329	DW	1	Mission Europe	LL	DWinet	0925	1525
1327	1351	CRI-ENG	1	People in the Know	GL	9570, 11885, 15230	0927	2127
1327	1351	CRI-ENG	2	Biz China	BE	9570, 11885, 15230	0927	2127
1327	1351	CRI-ENG	3	China Horizons	DL	9570, 11885, 15230	0927	2127
1327	1351	CRI-ENG	4	Voices from Other Lands	DL	9570, 11885, 15230	0927	2127
1327	1351	CRI-ENG	5	Life in China	DL	9570, 11885, 15230	0927	2127
1327	1351	CRI-ENG	6	Listener's Garden	LI	9570, 11885, 15230	0927	2127
1327	1351	CRI-ENG	7	In the Spotlight	DL	9570, 11885, 15230	0927	2127
1330	1359	ABC-RA	2	Innovations	ST	RAinet	0930	2330
1330	1359	ABC-RA	3	Australian Bite	DL	RAinet	0930	2330
1330	1359	ABC-RA	4	Rural Reporter	DL	RAinet	0930	2330
1330	1359	ABC-RA	5	Rear Vision	NA	RAinet	0930	2330
1330	1359	ABC-RA	6	All in the Mind	HM	RAinet	0930	2330
1330	1429	BBC-R4	7	The Saturday Play	LD	BBC-R4inet	0930	1430
1330	1359	BBCWS-AM	1	The Interview	PI	XM131	0930	1430
1330	1359	BBCWS-AM	2	Health Check	HM	XM131	0930	1430
1330	1359	BBCWS-AM	3	Digital Planet	DX	XM131	0930	1430
1330	1359	BBCWS-AM	4	Discovery	ST	XM131	0930	1430
1330	1359	BBCWS-AM	5	One Planet	EV	XM131	0930	1430
1330	1359	BBCWS-AM	6	Science in Action	ST	XM131	0930	1430
1330	1359	BBCWS-IE	7	Charlie Gillett's World of Music	NZ	BBCWS-IEinet	0900	1400
1330	1359	BBCWS-NX	1	Reporting Religion	CS	BBCWS-NXinet	0930	1430
1330	1359	BBCWS-NX	7	The Interview	PI	BBCWS-NXinet	0930	1430
1330	1359	BBCWS-PR	1	Reporting Religion	CS	Sirius141, NPRfm/am, HD+	0930	1430
1330	1359	BBCWS-PR	7	The Interview	PI	Sirius141, NPRfm/am, HD+	0930	1430
1330	1459	CBC-R1C	2-6	The Current	NA	CBWinet, 990	0930	0830
1330	1359	CRI-RTC	1	China Horizons	DL	CRI-RTCinet	0930	2130
1330	1359	CRI-RTC	2	Frontline	GL	CRI-RTCinet	0930	2130
1330	1359	CRI-RTC	3	Biz China	BE	CRI-RTCinet	0930	2130
1330	1359	CRI-RTC	4	In the Spotlight	DL	CRI-RTCinet	0930	2130
1330	1359	CRI-RTC	5	Voices from Other Lands	DL	CRI-RTCinet	0930	2130
1330	1359	CRI-RTC	6	Life in China	DL	CRI-RTCinet	0930	2130
1330	1359	CRI-RTC	7	Listeners' Garden	LI	CRI-RTCinet	0930	2130
1330	1359	CRI-WASH	1	China Horizons	DL	CRI-WASHinet	0930	2130
1330	1359	CRI-WASH	2	Listener's Garden	LI	CRI-WASHinet	0930	2130
1330	1359	CRI-WASH	3	Frontline	GL	CRI-WASHinet	0930	2130
1330	1359	CRI-WASH	4	Biz China	BE	CRI-WASHinet	0930	2130
1330	1359	CRI-WASH	5	In the Spotlight	DL	CRI-WASHinet	0930	2130
1330	1359	CRI-WASH	6	Voices from Other Lands	DL	CRI-WASHinet	0930	2130
1330	1359	CRI-WASH	7	Life in China	DL	CRI-WASHinet	0930	2130
1330	1359	DW	1	A World of Music	MV	DWinet	0930	1530
1330	1359	DW	2	Eurovox	AC	DWinet	0930	1530
1330	1359	DW	3	Hits in Germany	MP	DWinet	0930	1530
1330	1359	DW	4	Arts on the Air	AC	DWinet	0930	1530
1330	1359	DW	5	Cool	CS	DWinet	0930	1530
1330	1359	DW	6	Dialogue	CS	DWinet	0930	1530
1330	1359	RNW	1	Reloaded	VA	RNW2inet	0930	1530
1330	1359	RNW	2	Curious Orange	DL	RNW2inet	0930	1530
1330	1359	RNW	3	The State We're In Midweek Edition	CS	RNW2inet	0930	1530
1330	1359	RNW	4	Radio Books	LD	RNW2inet	0930	1530
1330	1359	RNW	5	Earthbeat	EV	RNW2inet	0930	1530
1330	1359	RNW	6	Bridges with Africa	CS	RNW2inet	0930	1530
1330	1359	RNW	7	Curious Orange	DL	RNW2inet	0930	1530
1330	1359	RNZI	2	Mailbox (fortnightly)	DX	RNZIinet, 6170s, 13840w	0930	0130

UTC Time Start	End	Station/ Network	Day(s)	Program Name	Type	Frequency/ Platform	EDT	Station Time
1330	1359	RNZI	2	Spectrum (fortnightly)	DL	RNZIinet, 6170s, 13840w	0930	0130
1330	1359	RNZI	3	Tradewinds	BE	RNZIinet, 6170s, 13840w	0930	0130
1330	1359	RNZI	4	World in Sport	SP	RNZIinet, 6170s, 13840w	0930	0130
1330	1359	RNZI	5	Pacific Correspondent	NA	RNZIinet, 6170s, 13840w	0930	0130
1330	1359	RNZI	6	Waiata	MF	RNZIinet, 6170s, 13840w	0930	0130
1345	1359	BBC-R4	1	Music documentary or feature series	MD	BBC-R4inet	0945	1445
1352	1357	CRI-ENG	1-7	Learning Chinese Now	LL	9570, 11885, 15230	0952	2152
1400	1459	ABC-RA	2	Big Ideas	ED	RAinet	1000	0000
1400	1459	ABC-RA	3	AWAYE!	DL	RAinet	1000	0000
1400	1429	ABC-RA	4	All in the Mind	HM	RAinet	1000	0000
1400	1459	ABC-RA	5	Hindsight	AC	RAinet	1000	0000
1400	1429	ABC-RA	6	MovieTime	AC	RAinet	1000	0000
1400	1444	ABC-RN	1	The Book Show (from 2/09)	AC	ABC-RNinet	1000	0000
1400	1444	ABC-RN	1-5	The Book Show (til 1/09)	AC	ABC-RNinet	1000	0000
1400	1429	ABC-RN	2	Health Report (from 2/09)	HM	ABC-RNinet	1000	0000
1400	1429	ABC-RN	3	Law Report (from 2/09)	GL	ABC-RNinet	1000	0000
1400	1429	ABC-RN	4	Rear Vision (from 2/09)	NA	ABC-RNinet	1000	0000
1400	1429	ABC-RN	5	Future Report (from 2/09)	ST	ABC-RNinet	1000	0000
1400	1429	ABC-RN	6	Sound Quality (from 2/09)	MZ	ABC-RNinet	1000	0000
1400	1459	ABC-RN	6	The Night Air (til 1/09)	LD	ABC-RNinet	1000	0000
1400	1459	BBC-R4	1	The Classic Serial	LD	BBC-R4inet	1000	1500
1400	1429	BBC-R4	2	Money Box Live	BE	BBC-R4inet	1000	1500
1400	1429	BBC-R4	3	Making History	AC	BBC-R4inet	1000	1500
1400	1429	BBC-R4	4	Gardeners' Question Time	DL	BBC-R4inet	1000	1500
1400	1426	BBC-R4	5	Questions, Questions	ED	BBC-R4inet	1000	1500
1400	1429	BBC-R4	6	Ramblings	DL	BBC-R4inet	1000	1500
1400	1459	BBCWS-AM	1	The Forum	CA	XM131	1000	1500
1400	1429	BBCWS-AM	2/4/6	Documentary feature or series	ND	XM131	1000	1500
1400	1429	BBCWS-AM	3	Global Business	BE	XM131	1000	1500
1400	1429	BBCWS-AM	5	Assignment	ND	XM131	1000	1500
1400	1659	BBCWS-AM	7	Sportsworld	SP	XM131	1000	1500
1400	1459	BBCWS-IE	1	The Strand	AC	BBCWS-IEinet	1000	1500
1400	1429	BBCWS-IE	2-6	Outlook	GZ	BBCWS-IEinet	1000	1500
1400	1659	BBCWS-IE	7	Sportsworld	SP	BBCWS-IEinet	1000	1500
1400	1429	BBCWS-NX	1	From Our Own Correspondent	NA	BBCWS-NXinet	1000	1500
1400	1439	BBCWS-NX	2-6	World Briefing	NX	BBCWS-NXinet	1000	1500
1400	1659	BBCWS-NX	7	Sportsworld	SP	BBCWS-NXinet	1000	1500
1400	1429	BBCWS-PR	1	From Our Own Correspondent	NA	Sirius141, NPRfm/am, HD+	1000	1500
1400	1429	BBCWS-PR	2-6	World Briefing	NX	Sirius141, NPRfm/am, HD+	1000	1500
1400	1459	BBCWS-PR	7	Sportsworld	SP	Sirius141, NPRfm/am, HD+	1000	1500
1400	1429	CBC-R1A	2	White Coat, Black Art	HM	CBAMinet, CBCTinet, CBDinet, CBHAinet, CBIinet, CBNinet, 640, 6160, CBTinet, CBYinet, CBZFinet, CFGBinet	1000	1100
1400	1429	CBC-R1A	3	C'est la vie	DL	CBAMinet, CBCTinet, CBDinet, CBHAinet, CBIinet, CBNinet, 640, 6160, CBTinet, CBYinet, CBZFinet, CFGBinet	1000	1100
1400	1429	CBC-R1A	4	Spark	DX	CBAMinet, CBCTinet, CBDinet, CBHAinet, CBIinet, CBNinet, 640, 6160, CBTinet, CBYinet, CBZFinet, CFGBinet	1000	1100
1400	1429	CBC-R1A	5	Afghanada	LD	CBAMinet, CBCTinet, CBDinet, CBHAinet, CBIinet, CBNinet, 640, 6160, CBTinet, CBYinet, CBZFinet, CFGBinet	1000	1100
1400	1659	CBC-R1C	1	The Sunday Edition	CA	CBWinet, 990	1000	0900
1400	1459	CBC-R1C	7	The House	GL	CBWinet, 990	1000	0900
1400	1459	CBC-R1E	2-5	Q	AC	CBCLinet, CBCSinet, CBEinet, 1550, CBLAinet, CBMEinet, CBOinet, CBQTinet, CBVEinet, CFFBinet	1000	1000
1400	1529	CBC-R1E	6	Q	AC	CBCLinet, CBCSinet, CBEinet, 1550, CBLAinet, CBMEinet, CBOinet, CBQTinet, CBVEinet, CFFBinet	1000	1000
1400	1529	CBC-R1E	7	Go!	LE	CBCLinet, CBCSinet, CBEinet, 1550, CBLAinet, CBMEinet, CBOinet, CBQTinet, CBVEinet, CFFBinet	1000	1000
1400	1459	CBC-R1S	1	Tapestry	CS	Sirius137	1000	1000
1400	1426	CRI-ENG	1-5	News and Reports	NZ	13675, 13740, 15230	1000	2200
1400	1419	CRI-ENG	6/7	News and Reports	NZ	13675, 13740, 15230	1000	2200
1400	1429	CRI-RTC	1-7	News and Reports	NZ	CRI-RTCinet	1000	2200
1400	1459	CRI-WASH	1/2	China Beat	MP	CRI-WASHinet	1000	2200
1400	1419	CRI-WASH	3-7	People in the Know	GL	CRI-WASHinet	1000	2200
1400	1405	DW	1-7	News	NX	DWinet	1000	1600
1400	1409	NHK-RJ	1-7	News	NX	11705	1000	2300
1400	1404	R.PRG	1	News	NX	R.PRGinet, 13580	1000	1600
1400	1409	R.PRG	2-6	News and Current Affairs	NX	R.PRGinet, 13580	1000	1600
1400	1414	RFI	1/7	International News	NX	RFImultilingual2inet, Sirius140, WRN-NAinet	1000	1600
1400	1419	RFI	2-6	International News	NX	RFImultilingual2inet, Sirius140, WRN-NAinet	1000	1600
1400	1405	RNW	1/7	News	NX	RNW2inet	1000	1600
1400	1429	RNW	2	Earthbeat	EV	RNW2inet	1000	1600
1400	1429	RNW	4	Curious Orange	DL	RNW2inet	1000	1600
1400	1429	RNW	5	Network Europe Extra	AC	RNW2inet	1000	1600
1400	1429	RNW	6	Radio Books	LD	RNW2inet	1000	1600
1400	1759	RNZI	1	(see RNZ-NAT)		RNZIinet, 6170w, 7145s	1000	0200
1400	1559	RNZI	2-6	(see RNZ-NAT)		RNZIinet, 6170w	1000	0200
1400	1934	RNZI	7	(see RNZ-NAT)		RNZIinet, 6170w, 7145s	1000	0200
1400	1529	RTE-R1	2-6	Mooney	CA	RTE-R1inet	1000	1500

UTC Time Start	End	Station/ Network	Day(s)	Program Name	Type	Frequncy/ Platform	Station EDT	Time
1400	1759	RTHK-3	1	Sunday Late	MP	RTHK-3inet	1000	2200
1400	1759	RTHK-3	2-6	All the Way with Ray	MP	RTHK-3inet	1000	2200
1400	1759	RTHK-3	7	Cool Trax	MJ	RTHK-3inet	1000	2200
1400	1414	SABC-CHAF	1-7	News	NX	CHAFStudio1inet	1000	1600
1400	1559	SABC-SAFM	1	SAfm Sports Special	LD	SABC-SAFMinet	1000	1600
1400	1559	SABC-SAFM	2-6	PM Live	NZ	SABC-SAFMinet	1000	1600
1400	1459	WRN-NA	1-7	Radio France International features	VA	Sirius140, WRN-NAinet	1000	1500
1400	1429	WRS-SUI	2-6	BBC World Briefing	NX	WRS-SUIinet	1000	1600
1405	1429	DW	2-6	Newslink	NZ	DWinet	1005	1605
1405	1429	DW	7	Network Europe Week	NZ	DWinet	1005	1605
1405	1459	DW	1	Concert Hour	MC	DWinet	1005	1605
1405	1414	R.PRG	1	Mailbox	LI	R.PRGinet, 13580	1005	1605
1405	1414	R.PRG	7	Magazine	DL	R.PRGinet, 13580	1005	1605
1405	1429	RNW	1/3	Bridges with Africa	CS	RNW2inet	1005	1605
1405	1459	RNW	7	The State We're In	CS	RNW2inet	1005	1605
1405	1430	RNZ-NAT	2	BBC Feature	GI	RNZ-NATinet	1005	0205
1405	1430	RNZ-NAT	3	The Forum	CA	RNZ-NATinet	1005	0205
1405	1429	RNZ-NAT	4	Playing Favourites	MP	RNZ-NATinet	1005	0205
1406	1429	RNZ-NAT	7	Touchstone	CS	RNZ-NATinet	1006	0206
1410	1429	NHK-RJ	1	Pop Up Japan	MP	11705	1010	2310
1410	1424	NHK-RJ	2/4/6	What's Up Japan	NZ	11705	1010	2310
1410	1429	NHK-RJ	3/5	What's Up Japan	NZ	11705	1010	2310
1410	1429	NHK-RJ	7	World Interactive	LI	11705	1010	2310
1410	1419	R.PRG	2	One on One	PI	R.PRGinet, 13580	1010	1610
1410	1428	R.PRG	3	Talking Point	DL	R.PRGinet, 13580	1010	1610
1410	1428	R.PRG	4	Czechs in History (monthly)	AC	R.PRGinet, 13580	1010	1610
1410	1428	R.PRG	4	Czechs Today (monthly)	CS	R.PRGinet, 13580	1010	1610
1410	1428	R.PRG	4	Spotlight (fortnightly)	TR	R.PRGinet, 13580	1010	1610
1410	1418	R.PRG	5	Panorama	CS	R.PRGinet, 13580	1010	1610
1410	1418	R.PRG	6	Business News	BE	R.PRGinet, 13580	1010	1610
1415	1419	R.PRG	1	Letter from Prague	DL	R.PRGinet, 13580	1015	1615
1415	1418	R.PRG	7	Sound Czech	LL	R.PRGinet, 13580	1019	1619
1415	1419	RFI	7	Asia Pacific	NA	RFImultilingual2inet, Sirius140, WRN-NAinet	1015	1615
1415	1459	SABC-CHAF	1	Our Heritage	AC	CHAFStudio1inet	1015	1615
1415	1459	SABC-CHAF	2	African Music	MF	CHAFStudio1inet	1015	1615
1415	1459	SABC-CHAF	3-6	Current Affairs	NA	CHAFStudio1inet	1015	1615
1415	1459	SABC-CHAF	7	Variety Music	MV	CHAFStudio1inet	1015	1615
1419	1428	R.PRG	1	Czech Books (fortnightly)	AC	R.PRGinet, 13580	1019	1619
1419	1428	R.PRG	1	Magic Carpet (monthly)	MZ	R.PRGinet, 13580	1019	1619
1419	1428	R.PRG	1	Music Profile (monthly)	MX	R.PRGinet, 13580	1019	1619
1419	1428	R.PRG	6	The Arts	AC	R.PRGinet, 13580	1019	1619
1419	1428	R.PRG	7	One on One	PI	R.PRGinet, 13580	1019	1619
1420	1426	CRI-ENG	6	CRI Roundup	NX	13675, 13740, 15230	1020	2220
1420	1426	CRI-ENG	7	Reports from Developing Countries	BE	13675, 13740, 15230	1020	2220
1420	1439	CRI-WASH	3-7	China Horizons	DL	CRI-WASHinet	1020	2220
1420	1428	R.PRG	2	Sports News	SP	R.PRGinet, 13580	1020	1620
1420	1424	RFI	2-7	Today in France	DL	RFImultilingual2inet, Sirius140, WRN-NAinet	1020	1620
1425	1429	NHK-RJ	2/4/6	Easy Japanese	LL	11705	1025	2325
1425	1429	RFI	2-7	Sport	SP	RFImultilingual2inet, Sirius140, WRN-NAinet	1025	1625
1427	1429	BBC-R4	5	The Radio 4 Appeal	DL	BBC-R4inet	1027	1527
1427	1451	CRI-ENG	1	People in the Know	GL	13675, 13740, 15230	1027	2227
1427	1451	CRI-ENG	2	Biz China	BE	13675, 13740, 15230	1027	2227
1427	1451	CRI-ENG	3	China Horizons	DL	13675, 13740, 15230	1027	2227
1427	1451	CRI-ENG	4	Voices from Other Lands	DL	13675, 13740, 15230	1027	2227
1427	1451	CRI-ENG	5	Life in China	DL	13675, 13740, 15230	1027	2227
1427	1451	CRI-ENG	6	Listener's Garden	LI	13675, 13740, 15230	1027	2227
1427	1451	CRI-ENG	7	In the Spotlight	DL	13675, 13740, 15230	1027	2227
1430	1459	ABC-RA	4	The Philosopher's Zone	ED	RAinet	1030	0030
1430	1459	ABC-RA	6	Artworks	AC	RAinet	1030	0030
1430	1459	ABC-RN	2-5	Asia Pacific (from 2/09)	NZ	ABC-RNinet	1030	0030
1430	1444	BBC-R4	2-6	The Afternoon Reading	LD	BBC-R4inet	1030	1530
1430	1459	BBC-R4	7	Soul Music	MD	BBC-R4inet	1030	1530
1430	1459	BBCWS-AM	2	Health Check	HM	XM131	1030	1530
1430	1459	BBCWS-AM	3	Digital Planet	DX	XM131	1030	1530
1430	1459	BBCWS-AM	4	Discovery	ST	XM131	1030	1530
1430	1459	BBCWS-AM	5	One Planet	EV	XM131	1030	1530
1430	1459	BBCWS-AM	6	Science in Action	ST	XM131	1030	1530
1430	1459	BBCWS-IE	2-6	The Strand	AC	BBCWS-IEinet	1030	1530
1430	1459	BBCWS-NX	1	Politics UK	GL	BBCWS-NXinet	1030	1530
1430	1459	BBCWS-PR	1	Politics UK	GL	Sirius141, NPRfm/am, HD+	1030	1530
1430	1449	BBCWS-PR	2-6	Business Daily	BE	Sirius141, NPRfm/am, HD+	1030	1530
1430	1459	CBC-R1A	7	The Debaters	LE	CBAMinet, CBCTinet, CBDinet, CBHAinet, CBIinet, CBNinet, 640, 6160, CBTinet, CBYinet, CBZFinet, CFGBinet	1030	1130
1430	1559	CBC-R1M	2-6	The Current	NA	CBKinet, 540, CBRinet, 1010, CBXinet, 740, CFYKinet, CHAKinet	1030	0830
1430	1459	CBC-R1S	7	Laugh Out Loud	LE	Sirius137	1030	1030
1430	1459	CRI-RTC	1	China Horizons	DL	CRI-RTCinet	1030	2230

UTC Time Start	End	Station/ Network	Day(s)	Program Name	Type	Frequncy/ Platform	EDT	Station Time
1430	1459	CRI-RTC	2	Frontline	GL	CRI-RTCinet	1030	2230
1430	1459	CRI-RTC	3	Biz China	BE	CRI-RTCinet	1030	2230
1430	1459	CRI-RTC	4	In the Spotlight	DL	CRI-RTCinet	1030	2230
1430	1459	CRI-RTC	5	Voices from Other Lands	DL	CRI-RTCinet	1030	2230
1430	1459	CRI-RTC	6	Life in China	DL	CRI-RTCinet	1030	2230
1430	1459	CRI-RTC	7	Listeners' Garden	LI	CRI-RTCinet	1030	2230
1430	1459	DW	2	World in Progress	BE	DWinet	1030	1630
1430	1459	DW	3	Spectrum	ST	DWinet	1030	1630
1430	1459	DW	4	Money Talks	BE	DWinet	1030	1630
1430	1459	DW	5	Living Planet	EV	DWinet	1030	1630
1430	1459	DW	6	Inside Europe	NZ	DWinet	1030	1630
1430	1459	DW	7	Insight	NA	DWinet	1030	1630
1430	1459	RFI	1	Club 9516	LI	RFImultilingual2inet, Sirius140, WRN-NAinet	1030	1630
1430	1459	RFI	2	Network Europe Extra	AC	RFImultilingual2inet, Sirius140, WRN-NAinet	1030	1630
1430	1459	RFI	3	Crossroads	CS	RFImultilingual2inet, Sirius140, WRN-NAinet	1030	1630
1430	1459	RFI	4	Voices	PI	RFImultilingual2inet, Sirius140, WRN-NAinet	1030	1630
1430	1459	RFI	5	Rendezvous	CS	RFImultilingual2inet, Sirius140, WRN-NAinet	1030	1630
1430	1459	RFI	6	World Tracks	MZ	RFImultilingual2inet, Sirius140, WRN-NAinet	1030	1630
1430	1459	RFI	7	Mission Paris	LL	RFImultilingual2inet, Sirius140, WRN-NAinet	1030	1630
1430	1459	RNW	1	Radio Books	LD	RNW2inet	1030	1630
1430	1459	RNW	2-6	Newsline	NZ	RNW2inet	1030	1630
1430	1459	RNZ-NAT	1	Hidden Treasures with Trevor Reekie	MZ	RNZ-NATinet	1030	0230
1430	1459	RNZ-NAT	5	The Sampler	MP	RNZ-NATinet	1030	0230
1430	1459	RNZ-NAT	6	Waiata	MF	RNZ-NATinet	1030	0230
1430	1459	RNZ-NAT	6	Waiata	MF	RNZ-NATinet	1030	0230
1430	1659	WRS-SUI	2-6	Drive Time	NZ	WRS-SUInet	1030	1630
1431	1445	WRS-SUI	2-6	Switzerland Today	DL	WRS-SUInet	1031	1631
1437	1559	CBC-R1S		The Current	NA	Sirius137	1037	1037
1440	1449	BBCWS-NX	2-6	Analysis	NA	BBCWS-NXinet	1040	1540
1440	1459	CRI-WASH	3-7	Cultural Carousel	AC	CRI-WASHinet	1040	2240
1445	1459	ABC-RN	1	First Person (from 2/09)	LD	ABC-RNinet	1045	0045
1445	1459	ABC-RN	1-5	First Person (til 1/09)	LD	ABC-RNinet	1445	0045
1445	1459	BBC-R4	2-6	Documentary feature or series	ND	BBC-R4inet	1045	1545
1450	1459	BBCWS-NX	2-6	Sports Roundup	SP	BBCWS-NXinet	1050	1550
1450	1459	BBCWS-PR	2-6	Analysis	NA	Sirius141, NPRfm/am, HD+	1050	1550
1452	1457	CRI-ENG	1-7	Learning Chinese Now	LL	13675, 13740, 15230	1052	2252
1500	1529	ABC-RA	2-6	Asia Pacific	NZ	RAinet	1100	0100
1500	1559	ABC-RN	1	Artworks	AC	ABC-RNinet	1100	0100
1500	1529	ABC-RN	2-6	Asia Pacific (til 1/09)	NZ	ABC-RNinet	1100	0100
1500	1559	ABC-RN	2-6	Bush Telegraph (from 2/09)	DL	ABC-RNinet	1100	0100
1500	1559	ABC-RN	7	Saturday Extra - part 2	CA	ABC-RNinet	1100	0100
1500	1529	BBC-R4	1/5	Open Book	AC	BBC-R4inet	1100	1600
1500	1529	BBC-R4	2	The Food Programme	CS	BBC-R4inet	1100	1600
1500	1529	BBC-R4	3	Law in Action	GL	BBC-R4inet	1100	1600
1500	1529	BBC-R4	4	Thinking Allowed	ED	BBC-R4inet	1100	1600
1500	1529	BBC-R4	6	Last Word	CS	BBC-R4inet	1100	1600
1500	1559	BBC-R4	7	Weekend Woman's Hour	CS	BBC-R4inet	1100	1600
1500	1529	BBCWS-AM	1	Documentary feature or series	ND	XM131	1100	1600
1500	1539	BBCWS-AM	2-6	World Briefing	NX	XM131	1100	1600
1500	1659	BBCWS-IE	1	Sunday Sportsworld	SP	BBCWS-IEinet	1100	1600
1500	1519	BBCWS-IE	2-6	World Briefing	NX	BBCWS-IEinet	1100	1600
1500	1659	BBCWS-NX	1	Sportsworld	SP	BBCWS-NXinet	1100	1600
1500	1539	BBCWS-NX	2-6	World Briefing	NX	BBCWS-NXinet	1100	1600
1500	1559	BBCWS-PR	1	The Forum	CA	Sirius141, NPRfm/am, HD+	1100	1600
1500	1539	BBCWS-PR	2-6	World Briefing	NX	Sirius141, NPRfm/am, HD+	1100	1600
1500	1559	BBCWS-PR	7	The Strand	AC	Sirius141, NPRfm/am, HD+	1100	1600
1500	1559	CBC-R1A	1	The Vinyl Cafe	LE	CBAMinet, CBCTinet, CBDinet, CBHAinet, CBInet, CBNinet, 640, 6160, CBTinet, CBYinet, CBZFinet, CFGBinet	1100	1200
1500	1659	CBC-R1A	2-6	Local noon-hour program	NZ	CBAMinet, CBCTinet, CBDinet, CBHAinet, CBInet, CBNinet, 640, 6160, CBTinet, CBYinet, CBZFinet, CFGBinet	1100	1200
1500	1559	CBC-R1A	7	Quirks and Quarks	ST	CBAMinet, CBCTinet, CBDinet, CBHAinet, CBInet, CBNinet, 640, 6160, CBTinet, CBYinet, CBZFinet, CFGBinet	1100	1200
1500	1559	CBC-R1C	2-5	Q	AC	CBWinet, 990	1100	1000
1500	1629	CBC-R1C	6	Q	AC	CBWinet, 990	1100	1000
1500	1629	CBC-R1C	7	Go!	LE	CBWinet, 990	1100	1000
1500	1529	CBC-R1E	2	White Coat, Black Art	HM	CBCLinet, CBCSinet, CBEinet, 1550, CBLAinet, CBMEinet, CBOinet, CBQTinet, CBVEinet, CFFBinet	1100	1100
1500	1529	CBC-R1E	3	C'est la vie	DL	CBCLinet, CBCSinet, CBEinet, 1550, CBLAinet, CBMEinet, CBOinet, CBQTinet, CBVEinet, CFFBinet	1100	1100
1500	1529	CBC-R1E	4	Spark	DX	CBCLinet, CBCSinet, CBEinet, 1550, CBLAinet, CBMEinet, CBOinet, CBQTinet, CBVEinet, CFFBinet	1100	1100
1500	1529	CBC-R1E	5	Afghanada	LD	CBCLinet, CBCSinet, CBEinet, 1550, CBLAinet, CBMEinet, CBOinet, CBQTinet, CBVEinet, CFFBinet	1100	1100
1500	1759	CBC-R1M	1	The Sunday Edition	CA	CBKinet, 540, CBRinet, 1010, CBXinet, 740, CFYKinet, CHAKinet	1100	0900
1500	1559	CBC-R1M	7	The House	GL	CBKinet, 540, CBRinet, 1010, CBXinet, 740, CFYKinet, CHAKinet	1100	0900
1500	1559	CBC-R1S	1	Writers and Company	AC	Sirius137	1100	1100
1500	1659	CBC-R1S	7	Definitely Not the Opera	CS	Sirius137	1100	1100

UTC Time Start	End	Station/ Network	Day(s)	Program Name	Type	Frequency/ Platform	EDT	Station Time
1500	1659	CBC-RCI	1	The Maple Leaf Mailbag	LI	Sirius95	1100	1100
1500	1659	CBC-RCI	2-6	The Link	GZ	Sirius95	1100	1100
1500	1659	CBC-RCI	7	Masala Canada	GZ	Sirius95	1100	1100
1500	1526	CRI-ENG	1-5	News and Reports	NZ	13740	1100	2300
1500	1519	CRI-ENG	6/7	News and Reports	NZ	13740	1100	2300
1500	1529	CRI-RTC	1-7	News and Reports	NZ	CRI-RTCinet	1100	2300
1500	1549	CRI-WASH	1-7	China Drive	NZ	CRI-WASHinet	1100	2300
1500	1505	DW	1-7	News	NX	DWinet	1100	1700
1500	1509	KBS-WR	1-7	News	NX	KBS-WR1inet	1100	0000
1500	1659	ORF-FM4	1	World Wide Show (in German and English) 17000	MZ	ORF-FM4inet	1100	
1500	1529	RNW	1	Network Europe Extra	AC	RNW2inet	1100	1700
1500	1529	RNW	2-6	Network Europe	NZ	RNW2inet	1100	1700
1500	1529	RNW	7	Network Europe Week	NA	RNW2inet	1100	1700
1500	1529	RTHK-3	2-6	News at Eleven	NX	RTHK-3inet	1100	2300
1500	1509	RTHK-3	7	News at Eleven	NX	RTHK-3inet	1100	2300
1500	1514	SABC-CHAF	1-7	News	NX	CHAFStudio1inet	1100	1700
1500	1529	WRN-NA	1-6	Radio Romania International features	VA	Sirius140, WRN-NAinet	1100	1600
1500	1529	WRN-NA	7	World Vision	BE	Sirius140, WRN-NAinet	1100	1600
1505	1704	CBC-RCI	1	The Maple Leaf Mailbag	LI	9610, 9800, RCIinet	1105	1105
1505	1704	CBC-RCI	2-6	The Link	GZ	9610, 9800, RCIinet	1105	1105
1505	1704	CBC-RCI	7	Masala Canada	GZ	9610, 9800, RCIinet	1105	1105
1505	1524	DW	1	In-Box	MB	DWinet	1105	1705
1505	1529	DW	2-6	Newslink	NZ	DWinet	1105	1705
1505	1559	DW	7	Inside Europe	NZ	DWinet	1105	1705
1510	1529	KBS-WR	1	Korean Pop Interactive	MP	KBS-WR1inet	1110	0010
1510	1514	KBS-WR	2-6	News Commentary	NC	KBS-WR1inet	1110	0010
1510	1529	KBS-WR	7	Worldwide Friendship	LI	KBS-WR1inet	1110	0010
1515	1529	KBS-WR	2	Faces of Korea	CS	KBS-WR1inet	1115	0015
1515	1529	KBS-WR	3	Business Watch	BE	KBS-WR1inet	1115	0015
1515	1529	KBS-WR	4	Culture on the Move	AC	KBS-WR1inet	1115	0015
1515	1529	KBS-WR	5	Korea Today and Tomorrow	DL	KBS-WR1inet	1115	0015
1515	1529	KBS-WR	6	Seoul Report	DL	KBS-WR1inet	1115	0015
1515	1559	RNZ-NAT	1-7	Book reading or Short story	LD	RNZ-NATinet	1115	0315
1515	1559	SABC-CHAF	1/7	Africa This Week	NZ	CHAFStudio1inet	1115	1715
1515	1559	SABC-CHAF	2	African Music	MF	CHAFStudio1inet	1115	1715
1515	1559	SABC-CHAF	3-6	Current Affairs	NA	CHAFStudio1inet	1115	1715
1520	1529	BBCWS-IE	2-6	World Business Report	BE	BBCWS-IEinet	1120	1620
1520	1526	CRI-ENG	6	CRI Roundup	NX	13740	1120	2320
1520	1526	CRI-ENG	7	Reports from Developing Countries	BE	13740	1120	2320
1525	1529	DW	1	Mission Europe	LL	DWinet	1125	1725
1527	1551	CRI-ENG	1	People in the Know	GL	13740	1127	2327
1527	1551	CRI-ENG	2	Biz China	BE	13740	1127	2327
1527	1551	CRI-ENG	3	China Horizons	DL	13740	1127	2327
1527	1551	CRI-ENG	4	Voices from Other Lands	DL	13740	1127	2327
1527	1551	CRI-ENG	5	Life in China	DL	13740	1127	2327
1527	1551	CRI-ENG	6	Listener's Garden	LI	13740	1127	2327
1527	1551	CRI-ENG	7	In the Spotlight	DL	13740	1127	2327
1530	1559	ABC-RA	2	Health Report	HM	RAinet	1130	0130
1530	1559	ABC-RA	3	Law Report	GL	RAinet	1130	0130
1530	1559	ABC-RA	4	Rear Vision (til 2/09)	NA	RAinet	1130	0130
1530	1559	ABC-RA	4	Religion Report (til 1/09)	CS	RAinet	1130	0130
1530	1559	ABC-RA	5	Media Report (til 1/09)	DX	RAinet	1130	0130
1530	1559	ABC-RA	5	Futures Report (til 2/09)	ST	RAinet	1130	0130
1530	1559	ABC-RA	6	MovieTime (til 2/09)	AC	RAinet	1130	0130
1530	1559	ABC-RA	6	Sports Factor (til 1/09)	SP	RAinet	1130	0130
1530	1559	ABC-RN	2	Health Report (til 1/09)	HM	ABC-RNinet	1130	0130
1530	1559	ABC-RN	3	Law Report (til 1/09)	GL	ABC-RNinet	1130	0130
1530	1559	ABC-RN	4	Religion Report (til 1/09)	CS	ABC-RNinet	1130	0130
1530	1559	ABC-RN	5	Media Report (til 1/09)	DX	ABC-RNinet	1130	0130
1530	1559	ABC-RN	6	The Sports Factor (til 1/09)	SP	ABC-RNinet	1130	0130
1530	1559	BBC-R4	1	Poetry Please	LD	BBC-R4inet	1130	1630
1530	1559	BBC-R4	2	Traveller's Tree	TR	BBC-R4inet	1130	1630
1530	1559	BBC-R4	3	A Good Read	AC	BBC-R4inet	1130	1630
1530	1559	BBC-R4	4	Science or health series or feature	ST	BBC-R4inet	1130	1630
1530	1559	BBC-R4	5	The Material World	ST	BBC-R4inet	1130	1630
1530	1559	BBC-R4	6	The Film Programme	AC	BBC-R4inet	1130	1630
1530	1559	BBCWS-AM	1	One Planet	EV	XM131	1130	1630
1530	1559	BBCWS-IE	2	Health Check	HM	BBCWS-IEinet	1130	1630
1530	1559	BBCWS-IE	3	Digital Planet	DX	BBCWS-IEinet	1130	1630
1530	1559	BBCWS-IE	4	Discovery	ST	BBCWS-IEinet	1130	1630

UTC Time Start	End	Station/ Network	Day(s)	Program Name	Type	Frequncy/ Platform	EDT	Station Time
1530	1559	BBCWS-IE	5	One Planet	EV	BBCWS-IEinet	1130	1630
1530	1559	BBCWS-IE	6	Science in Action	ST	BBCWS-IEinet	1130	1630
1530	1559	CBC-R1E	7	The Debaters	LE	CBCLinet, CBCSinet, CBEinet, 1550, CBLAinet, CBMEinet, CBOinet, CBQTinet, CBVEinet, CFFBinet	1130	1130
1530	1659	CBC-R1P	2-6	The Current	NA	CBCVinet, CBTKinet, CBUinet, 690, 6160, CBYGinet, CFWHinet	1130	0830
1530	1559	CRI-RTC	1	China Horizons	DL	CRI-RTCinet	1130	2330
1530	1559	CRI-RTC	2	Frontline	GL	CRI-RTCinet	1130	2330
1530	1559	CRI-RTC	3	Biz China	BE	CRI-RTCinet	1130	2330
1530	1559	CRI-RTC	4	In the Spotlight	DL	CRI-RTCinet	1130	2330
1530	1559	CRI-RTC	5	Voices from Other Lands	DL	CRI-RTCinet	1130	2330
1530	1559	CRI-RTC	6	Life in China	DL	CRI-RTCinet	1130	2330
1530	1559	CRI-RTC	7	Listeners' Garden	LI	CRI-RTCinet	1130	2330
1530	1559	DW	1	A World of Music	MV	DWinet	1130	1730
1530	1559	DW	2	Eurovox	AC	DWinet	1130	1730
1530	1559	DW	3	Hits in Germany	MP	DWinet	1130	1730
1530	1559	DW	4	Arts on the Air	AC	DWinet	1130	1730
1530	1559	DW	5	Cool	CS	DWinet	1130	1730
1530	1559	DW	6	Dialogue	CS	DWinet	1130	1730
1530	1629	IRIB	1-7	Koran Reading, News and features	VA	IRIBinet	1130	1900
1530	1559	RNW	1/6	Bridges with Africa	CS	RNW2inet	1130	1730
1530	1559	RNW	2	Curious Orange	DL	RNW2inet	1130	1730
1530	1559	RNW	3	The State We're In Midweek Edition	CS	RNW2inet	1130	1730
1530	1559	RNW	4	Radio Books	LD	RNW2inet	1130	1730
1530	1559	RNW	5/7	Earthbeat	EV	RNW2inet	1130	1730
1530	1559	RNZ-NAT	2	New Zealand Books	AC	RNZ-NATinet	1130	0330
1530	1559	RNZ-NAT	3	An Author's View	AC	RNZ-NATinet	1130	0330
1530	1559	RNZ-NAT	4	The Word	AC	RNZ-NATinet	1130	0330
1530	1559	RNZ-NAT	5	Canterbury Tales	DL	RNZ-NATinet	1130	0330
1530	1729	RTE-R1	2-6	Drivetime	NZ	RTE-R1inet	1130	1630
1530	1559	WRN-NA	1-7	Radio Korea International features	VA	Sirius140, WRN-NAinet	1130	1630
1530	1544	WRS-SUI	2-6	Switzerland Today	DL	WRS-SUIinet	1130	1730
1535	1559	RNZ-NAT	1	New Zealand Society	CS	RNZ-NATinet	1135	0335
1535	1559	RNZ-NAT	6	The Week	NA	RNZ-NATinet	1135	0335
1535	1559	RNZ-NAT	7	Hymns for Sunday Morning	CS	RNZ-NATinet	1135	0335
1540	1549	BBCWS-AM	2-6	World Business Report	BE	XM131	1140	1640
1540	1549	BBCWS-NX	2-6	World Business Report	BE	BBCWS-NXinet	1140	1640
1540	1549	BBCWS-PR	2-6	Analysis	NA	Sirius141, NPRfm/am, HD+	1140	1640
1550	1559	BBCWS-AM	2-6	Sports Report	SP	XM131	1150	1650
1550	1559	BBCWS-NX	2-6	Sports Roundup	SP	BBCWS-NXinet	1150	1650
1550	1559	BBCWS-PR	2-6	World Business Report	BE	Sirius141, NPRfm/am, HD+	1150	1650
1550	1559	CRI-WASH	1-7	Chinese Melody	MF	CRI-WASHinet	1100	2300
1552	1557	CRI-ENG	1-7	Learning Chinese Now	LL	13740	1152	2352
1600	1659	ABC-RA	1	The Science Show	ST	RAinet	1200	0200
1600	1659	ABC-RA	2-5	Australia Talks	DL	RAinet	1200	0200
1600	1659	ABC-RA	6	National Interest	NA	RAinet	1200	0200
1600	1659	ABC-RA	7	The Margaret Throsby Interview	AC	RAinet	1200	0200
1600	1659	ABC-RN	1	New Dimensions	CS	ABC-RNinet	1200	0200
1600	1634	ABC-RN	2	All in the Mind	HM	ABC-RNinet	1200	0200
1600	1659	ABC-RN	3	The Spirit of Things	CS	ABC-RNinet	1200	0200
1600	1659	ABC-RN	4	Radio Eye	ND	ABC-RNinet	1200	0200
1600	1659	ABC-RN	5	Hindsight	AC	ABC-RNinet	1200	0200
1600	1659	ABC-RN	6	Counterpoint	CA	ABC-RNinet	1200	0200
1600	1659	ABC-RN	7	Quiet Space	MZ	ABC-RNinet	1200	0200
1600	1639	BBC-R4	1	File on 4	ND	BBC-R4inet	1200	1700
1600	1659	BBC-R4	2-6	PM	NZ	BBC-R4inet	1200	1700
1600	1629	BBC-R4	7	PM	NZ	BBC-R4inet	1200	1700
1600	1759	BBCWS-AM	1	Sunday Sportsworld	AC	XM131	1200	1700
1600	1659	BBCWS-AM	2-6	Europe Today	NZ	XM131	1200	1700
1600	1659	BBCWS-IE	2-6	Europe Today	NZ	BBCWS-IEinet	1200	1700
1600	1659	BBCWS-PR	1	The Strand	AC	Sirius141, NPRfm/am, HD+	1200	1700
1600	1659	BBCWS-PR	2-6	Europe Today	NZ	Sirius141, NPRfm/am, HD+	1200	1700
1600	1659	BBCWS-PR	2-6	Europe Today	NZ	BBCWS-NXinet	1200	1700
1600	1629	BBCWS-PR	7	Assignment	ND	Sirius141, NPRfm/am, HD+	1200	1700
1600	1629	CBC-R1A	1	WireTap	LE	CBAMinet, CBCTinet, CBDinet, CBHAinet, CBIinet, CBNinet, 640, 6160, CBTinet, CBYinet, CBZFinet, CFGBinet	1200	1300
1600	1759	CBC-R1A	7	Definitely Not the Opera	CS	CBAMinet, CBCTinet, CBDinet, CBHAinet, CBIinet, CBNinet, 640, 6160, CBTinet, CBYinet, CBZFinet, CFGBinet	1200	1300
1600	1629	CBC-R1C	2	White Coat, Black Art	HM	CBWinet, 990	1200	1100
1600	1629	CBC-R1C	3	C'est la vie	DL	CBWinet, 990	1200	1100
1600	1629	CBC-R1C	4	Spark	DX	CBWinet, 990	1200	1100
1600	1629	CBC-R1C	5	Afghanada	LD	CBWinet, 990	1200	1100
1600	1659	CBC-R1E	1	The Vinyl Cafe	LE	CBCLinet, CBCSinet, CBEinet, 1550, CBLAinet, CBMEinet, CBOinet, CBQTinet, CBVEinet, CFFBinet	1200	1200
1600	1759	CBC-R1E	2-6	Local noon-hour program	NZ	CBCLinet, CBCSinet, CBEinet, 1550, CBLAinet, CBMEinet, CBOinet, CBQTinet, CBVEinet, CFFBinet	1200	1200

UTC Time Start	End	Station/ Network	Day(s)	Program Name	Type	Frequncy/ Platform	Station EDT	Time
1600	1659	CBC-R1E	7	Quirks and Quarks	ST	CBCLinet, CBCSinet, CBEinet, 1550, CBLAinet, CBMEinet, CBOinet, CBQTinet, CBVEinet, CFFBinet	1200	1200
1600	1659	CBC-R1M	2-5	Q	AC	CBKinet, 540, CBRinet, 1010, CBXinet, 740, CFYKinet, CHAKinet	1200	1000
1600	1729	CBC-R1M	6	Q	AC	CBKinet, 540, CBRinet, 1010, CBXinet, 740, CFYKinet, CHAKinet	1200	1000
1600	1729	CBC-R1M	7	Go!	LE	CBKinet, 540, CBRinet, 1010, CBXinet, 740, CFYKinet, CHAKinet	1200	1000
1600	1859	CBC-R1P	1	The Sunday Edition	CA	CBCVinet, CBTKinet, CBUinet, 690, 6160, CBYGinet, CFWHinet	1200	0900
1600	1659	CBC-R1P	7	The House	GL	CBCVinet, CBTKinet, CBUinet, 690, 6160, CBYGinet, CFWHinet	1200	0900
1600	1612	CBC-R1S	1	World Report	NX	Sirius137	1200	1200
1600	1729	CBC-R1S	2-6	The Point	CA	Sirius137	1200	1200
1600	1629	CRI-RTC	1-7	News and Reports	NZ	CRI-RTCinet	1200	0000
1600	1629	CRI-WASH	1-7	Realtime China	NZ	CRI-WASHinet	1200	0000
1600	1605	DW	1-7	News	NX	DWinet	1200	1800
1600	1604	RNW	1/7	News	NX	RNW2inet	1200	1800
1600	1629	RNW	2-6	Network Europe	NZ	RNW2inet	1200	1800
1600	1607	RNZI	2-6	Pacific Regional News	NX	RNZIinet, 6170w, 7145s	1200	0400
1600	1659	SABC-SAFM	1	Living Sounds	MJ	SABC-SAFMinet	1200	1800
1600	1629	SABC-SAFM	2-6	Market Update	BE	SABC-SAFMinet	1200	1800
1600	1615	WRN-NA	1	UN Radio: UN Calling Asia	NZ	Sirius140, WRN-NAinet, 9955 (M-F)	1200	1700
1600	1615	WRN-NA	2-6	Radio New Zealand Int.: Korero Pacifica	NX	Sirius140, WRN-NAinet, 9955 (M-F)	1200	1700
1600	1615	WRN-NA	7	Radio Guangdong: Guangdong Today	DL	Sirius140, WRN-NAinet, 9955 (M-F)	1200	1700
1605	1629	DW	1-7	Newslink	NZ	DWinet	1205	1805
1605	1629	RNW	1	Network Europe Extra	AC	RNW2inet	1205	1805
1605	1659	RNW	7	The State We're In	CS	RNW2inet	1205	1805
1608	1629	RNZI	2-6	Dateline Pacific	NA	RNZIinet, 6170w, 7145s	1208	0408
1613	1859	CBC-R1S	1	The Sunday Edition	CS	Sirius137	1213	1213
1615	1629	WRN-NA	1-7	Vatican Radio: News	NX	Sirius140, WRN-NAinet, 9955 (M-F)	1215	1715
1630	1653	BBC-R4	7	The Bottom Line	BE	BBC-R4inet	1230	1730
1630	1659	BBCWS-PR	7	Business Weekly	BE	Sirius141, NPRfm/am, HD+	1230	1730
1630	1659	CBC-R1A	1	The Inside Track	SP	CBAMinet, CBCTinet, CBDinet, CBHAinet, CBIinet, CBNinet, 640, 6160, CBTinet, CBYinet, CBZFinet, CFGBinet	1230	1330
1630	1659	CBC-R1C	7	The Debaters	LE	CBWinet, 990	1230	1130
1630	1654	CRI-RTC	1	China Horizons	DL	CRI-RTCinet	1230	0030
1630	1654	CRI-RTC	2	Frontline	GL	CRI-RTCinet	1230	0030
1630	1654	CRI-RTC	3	Biz China	BE	CRI-RTCinet	1230	0030
1630	1654	CRI-RTC	4	In the Spotlight	DL	CRI-RTCinet	1230	0030
1630	1654	CRI-RTC	5	Voices from Other Lands	DL	CRI-RTCinet	1230	0030
1630	1654	CRI-RTC	6	Life in China	DL	CRI-RTCinet	1230	0030
1630	1654	CRI-RTC	7	Listeners' Garden	LI	CRI-RTCinet	1230	0030
1630	1659	CRI-WASH	1-7	News and Reports	NZ	CRI-WASHinet	1230	0030
1630	1659	DW	1	Insight	NA	DWinet	1230	1830
1630	1659	DW	2	World in Progress	BE	DWinet	1230	1830
1630	1659	DW	3	Spectrum	ST	DWinet	1230	1830
1630	1659	DW	4	Money Talks	BE	DWinet	1230	1830
1630	1659	DW	5	Living Planet	EV	DWinet	1230	1830
1630	1659	DW	6	Inside Europe	NZ	DWinet	1230	1830
1630	1659	DW	7	Dialogue	CS	DWinet	1230	1830
1630	1659	RNW	1	Reloaded	VA	RNW2inet	1230	1830
1630	1659	RNW	2	Earthbeat	EV	RNW2inet	1230	1830
1630	1659	RNW	3	Bridges with Africa	CS	RNW2inet	1230	1830
1630	1659	RNW	4	Curious Orange	DL	RNW2inet	1230	1830
1630	1659	RNW	5	Network Europe Extra	AC	RNW2inet	1230	1830
1630	1659	RNW	6	Radio Books	LD	RNW2inet	1230	1830
1630	1659	RNZ-NAT	5	Global Business	BE	RNZ-NATinet	1230	0430
1630	1659	RNZ-NAT	7	On Screen	AC	RNZ-NATinet	1230	0430
1630	1659	RNZI	2	Mailbox (fortnightly)	DX	RNZIinet, 6170w, 7145s	1330	0430
1630	1659	RNZI	2	Spectrum (fortnightly)	DL	RNZIinet, 6170w, 7145s	1230	0430
1630	1659	RNZI	3	Tradewinds	BE	RNZIinet, 6170w, 7145s	1230	0430
1630	1659	RNZI	4/6	Waiata	MF	RNZIinet, 6170w, 7145s	1230	0430
1630	1659	RNZI	5	Pacific Correspondent	NA	RNZIinet, 6170w, 7145s	1230	0430
1630	1659	SABC-SAFM	2-6	Gameplan	SP	SABC-SAFMinet	1230	1830
1630	1659	WRN-NA	1-7	Radio Slovakia International features	VA	Sirius140, WRN-NAinet, 9955 (M-F)	1230	1730
1635	1659	ABC-RN	2	The Philosopher's Zone	ED	ABC-RNinet	1235	0235
1640	1653	BBC-R4	1	From Fact to Fiction	LD	BBC-R4inet	1240	1740
1653	1659	RTE-R1	1	Weather Forecast and Nuacht	WX	RTE-R1inet	1253	1753
1654	1656	BBC-R4	1/7	Shipping Forecast	WX	BBC-R4inet	1254	1754
1655	1659	CRI-RTC	1-7	Chinese Studio	LL	CRI-RTCinet	1230	0030
1657	1659	BBC-R4	1/7	Weather	WX	BBC-R4inet	1257	1757
1700	1733	ABC-RA	1	Total Rugby	SP	RAinet	1300	0300
1700	1729	ABC-RA	2	Innovations	ST	RAinet	1300	0300
1700	1729	ABC-RA	3	Australian Bite	DL	RAinet	1300	0300
1700	1729	ABC-RA	4	Rural Reporter	DL	RAinet	1300	0300
1700	1729	ABC-RA	5	Rear Vision	NA	RAinet	1300	0300
1700	1729	ABC-RA	6	All in the Mind	HM	RAinet	1300	0300
1700	1759	ABC-RA	6	Big Ideas	ED	RAinet	1300	0300
1700	1759	ABC-RA	7	Late Night Live Classic	CA	RAinet	1300	0300
1700	1759	ABC-RN	1	Sound Quality	MZ	ABC-RNinet	1300	0300
1700	1754	ABC-RN	2-5	Australia Talks	DL	ABC-RNinet	1300	0300

UTC Time Start	End	Station/ Network	Day(s)	Program Name	Type	Frequency/ Platform	EDT	Station Time
1700	1754	ABC-RN	6	The National Interest	CA	ABC-RNinet	1300	0300
1700	1759	ABC-RN	7	AWAYE!	DL	ABC-RNinet	1300	0300
1700	1714	BBC-R4	1/7	The Six O'Clock News	NX	BBC-R4inet	1300	1800
1700	1729	BBC-R4	2-6	The Six O'Clock News	NX	BBC-R4inet	1300	1800
1700	1759	BBCWS-AM	2-6	World, Have Your Say	LI	XM131	1300	1800
1700	1729	BBCWS-AM	7	World Briefing	NX	XM131	1300	1800
1700	1729	BBCWS-IE	1	Documentary feature or series	ND	BBCWS-IEinet	1300	1800
1700	1759	BBCWS-IE	2-6	World, Have Your Say	LI	BBCWS-IEinet	1300	1800
1700	1729	BBCWS-IE	7	World Briefing	NX	BBCWS-IEinet	1300	1800
1700	1729	BBCWS-NX	1/7	World Briefing	NX	BBCWS-NXinet	1300	1800
1700	1759	BBCWS-NX	2-6	World, Have Your Say	LI	BBCWS-NXinet	1300	1800
1700	1729	BBCWS-PR	1/7	World Briefing	NX	Sirius141, NPRfm/am, HD+	1300	1800
1700	1759	BBCWS-PR	2-6	World, Have Your Say	LI	Sirius141, NPRfm/am, HD+	1300	1800
1700	1759	CBC-R1A	1	Tapestry	CS	CBAMinet, CBCTinet, CBDinet, CBHAinet, CBIinet, CBNinet, 640, 6160, CBTinet, CBYinet, CBZFinet, CFGBinet	1300	1400
1700	1759	CBC-R1A	2-6	The Point	CA	CBAMinet, CBCTinet, CBDinet, CBHAinet, CBIinet, CBNinet, 640, 6160, CBTinet, CBYinet, CBZFinet, CFGBinet	1300	1400
1700	1759	CBC-R1C	1	The Vinyl Cafe	LE	CBWinet, 990	1300	1200
1700	1859	CBC-R1C	2-6	Local noon-hour program	NZ	CBWinet, 990	1300	1200
1700	1759	CBC-R1C	7	Quirks and Quarks	ST	CBWinet, 990	1300	1200
1700	1729	CBC-R1E	1	WireTap	LE	CBCLinet, CBCSinet, CBEinet, 1550, CBLAinet, CBMEinet, CBOinet, CBQTinet, CBVEinet, CFFBinet	1300	1300
1700	1859	CBC-R1E	7	Definitely Not the Opera	CS	CBCLinet, CBCSinet, CBEinet, 1550, CBLAinet, CBMEinet, CBOinet, CBQTinet, CBVEinet, CFFBinet	1300	1300
1700	1729	CBC-R1M	2	White Coat, Black Art	HM	CBKinet, 540, CBRinet, 1010, CBXinet, 740, CFYKinet, CHAKinet	1300	1100
1700	1729	CBC-R1M	3	C'est la vie	DL	CBKinet, 540, CBRinet, 1010, CBXinet, 740, CFYKinet, CHAKinet	1300	1100
1700	1729	CBC-R1M	4	Spark	DX	CBKinet, 540, CBRinet, 1010, CBXinet, 740, CFYKinet, CHAKinet	1300	1100
1700	1729	CBC-R1M	5	Afghanada	LD	CBKinet, 540, CBRinet, 1010, CBXinet, 740, CFYKinet, CHAKinet	1300	1100
1700	1759	CBC-R1P	2-5	Q	AC	CBCVinet, CBTKinet, CBUinet, 690, 6160, CBYGinet, CFWHinet	1300	1000
1700	1829	CBC-R1P	6	Q	AC	CBCVinet, CBTKinet, CBUinet, 690, 6160, CBYGinet, CFWHinet	1300	1000
1700	1829	CBC-R1P	7	Go!	LE	CBCVinet, CBTKinet, CBUinet, 690, 6160, CBYGinet, CFWHinet	1300	1000
1700	1759	CBC-R1S	7	The Next Chapter	AC	Sirius137	1300	1300
1700	1719	CRI-RTC	1/7	News and Reports	NZ	CRI-RTCinet	1300	0100
1700	1754	CRI-RTC	2-6	China Drive	GZ	CRI-RTCinet	1300	0100
1700	1959	CRI-WASH	1-7	Beyond Beijing	MV	CRI-WASHinet	1300	0100
1700	1705	DW	1-7	News	NX	DWinet	1300	1900
1700	1724	PR-EXT	1	Network Europe Extra	AC	PR-EXTinet	1300	1900
1700	1719	PR-EXT	2-6	News from Poland	NX	PR-EXTinet	1300	1900
1700	1719	PR-EXT	7	Europe East	NZ	PR-EXTinet	1300	1900
1700	1704	R.PRG	1/7	News	NX	R.PRGinet	1300	1900
1700	1709	R.PRG	2-6	News and Current Affairs	NX	R.PRGinet	1300	1900
1700	1704	RNW	1/7	News	NX	RNW2inet	1300	1900
1700	1729	RNW	2-6	Newsline	NZ	RNW2inet	1300	1900
1700	1707	RNZI	2-6	Pacific Regional News	NX	RNZIinet, 6170w, 7145s	1300	0500
1700	1704	RTE-R1	1/7	The Angelus	CS	RTE-R1inet	1300	1800
1700	1714	SABC-CHAF		News	NX	CHAFStudio1inet	1300	1900
1700	1759	SABC-SAFM	1	Faith 2 Faith	CS	SABC-SAFMinet	1300	1900
1700	1859	SABC-SAFM	2-6	Evening Talk	NA	SABC-SAFMinet	1300	1900
1700	1759	SABC-SAFM	7	Drama	LD	SABC-SAFMinet	1300	1900
1700	1729	WRN-NA	1-7	Polish Radio External Service features	VA	Sirius140, WRN-NAinet, 9955 (M-F)	1300	1800
1700	1929	WRS-SUI	2-6	Relay BBC World Service	VA	WRS-SUIinet	1300	1900
1700	1759	WRS-SUI	7	Bookmark (2nd Sat. of the month)	AC	WRS-SUIinet	1300	1900
1705	1729	DW	1-7	Newslink	NZ	DWinet	1305	1905
1705	1714	R.PRG	1	Mailbox	LI	R.PRGinet	1305	1905
1705	1714	R.PRG	7	Magazine	DL	R.PRGinet	1305	1905
1705	1729	RNW	1	Network Europe Extra	AC	RNW2inet	1305	1905
1705	1759	RNW	7	The State We're In	CS	RNW2inet	1305	1905
1705	1759	RTE-R1	1	Spirit Moves	CS	RTE-R1inet	1305	1905
1705	1759	RTE-R1	7	Spectrum	DL	RTE-R1inet	1304	1804
1708	1729	RNZI	3	Tradewinds	BE	RNZIinet, 6170w, 7145s	1308	0508
1708	1729	RNZI	4	World in Sport	SP	RNZIinet, 6170w, 7145s	1308	0508
1708	1729	RNZI	5	Pacific Correspondent	NA	RNZIinet, 6170w, 7145s	1308	0508
1708	1749	RNZI	6	(see RNZ-NAT)		RNZIinet, 6170w, 7145s	1308	0508
1710	1719	R.PRG	2	One on One	PI	R.PRGinet	1310	1910
1710	1728	R.PRG	3	Talking Point	DL	R.PRGinet	1310	1910
1710	1728	R.PRG	4	Czechs in History (monthly)	AC	R.PRGinet	1310	1910
1710	1728	R.PRG	4	Czechs Today (monthly)	CS	R.PRGinet	1310	1910
1710	1728	R.PRG	4	Spotlight (fortnightly)	TR	R.PRGinet	1310	1910
1710	1718	R.PRG	5	Panorama	CS	R.PRGinet	1310	1910
1710	1718	R.PRG	6	Business News	BE	R.PRGinet	1310	1910
1710	1729	RNZ-NAT	1-5	He Rourou (in Maori)	DL	RNZ-NATinet	1310	0510
1710	1729	RNZ-NAT	7	Nga Marae	DL	RNZ-NATinet	1310	0510
1715	1759	BBC-R4	1	Pick of the Week	VA	BBC-R4inet	1315	1815
1715	1759	BBC-R4	7	Loose Ends	LE	BBC-R4inet	1315	1815

UTC Time Start	End	Station/ Network	Day(s)	Program Name	Type	Frequncy/ Platform	EDT	Station Time
1715	1719	R.PRG	1	Letter from Prague	DL	R.PRGinet	1315	1915
1715	1718	R.PRG	7	Sound Czech	LL	R.PRGinet	1319	1919
1715	1759	SABC-CHAF	1	SADC Calling	BE	CHAFStudio1inet	1315	1915
1715	1759	SABC-CHAF	2-6	Current Affairs	NA	CHAFStudio1inet	1315	1915
1715	1759	SABC-CHAF	7	Africa This Week	NA	CHAFStudio1inet	1315	1915
1719	1728	R.PRG	1	Czech Books (fortnightly)	AC	R.PRGinet	1319	1919
1719	1728	R.PRG	1	Magic Carpet (monthly)	MZ	R.PRGinet	1319	1919
1719	1728	R.PRG	1	Music Profile (monthly)	MX	R.PRGinet	1319	1919
1719	1728	R.PRG	6	The Arts	AC	R.PRGinet	1319	1919
1719	1728	R.PRG	7	One on One	PI	R.PRGinet	1319	1919
1720	1726	CRI-RTC	1	Reports from Developing Countries	BE	CRI-RTCinet	1320	0120
1720	1726	CRI-RTC	7	CRI Roundup	NX	CRI-RTCinet	1320	0120
1720	1729	PR-EXT	2	Around Poland	TR	PR-EXTinet	1320	1920
1720	1729	PR-EXT	3	Letter from Poland	DL	PR-EXTinet	1320	1920
1720	1734	PR-EXT	4	Day in the Life	PI	PR-EXTinet	1320	1920
1720	1724	PR-EXT	5	Comment	NC	PR-EXTinet	1320	1920
1720	1729	PR-EXT	6	Business Week	BE	PR-EXTinet	1320	1920
1720	1729	PR-EXT	7	A Look at the Weeklies	PR	PR-EXTinet	1320	1920
1720	1728	R.PRG	2	Sports News	SP	R.PRGinet	1320	1920
1725	1734	PR-EXT	1	The Kids	CS	PR-EXTinet	1325	1925
1725	1734	PR-EXT	5	Focus	AC	PR-EXTinet	1325	1925
1727	1754	CRI-RTC	1/7	China Beat	MP	CRI-RTCinet	1327	0127
1730	1759	ABC-RA	2-6	In the Loop	CA	RAinet	1330	0330
1730	1759	BBC-R4	2	Quiz or panel game	LE	BBC-R4inet	1330	1830
1730	1759	BBC-R4	3-5	Comedy series	LE	BBC-R4inet	1330	1830
1730	1759	BBC-R4	6	The News Quiz	LE	BBC-R4inet	1330	1830
1730	1759	BBCWS-AM	7	Sportsworld Extra	SP	XM131	1330	1830
1730	1759	BBCWS-IE	1	The Interview	PI	BBCWS-IEinet	1330	1830
1730	1759	BBCWS-IE	7	Sportsworld Extra	SP	BBCWS-IEinet	1330	1830
1730	1759	BBCWS-NX	1	The Interview	PI	BBCWS-NXinet	1330	1830
1730	1759	BBCWS-NX	7	Sportsworld Extra	SP	BBCWS-NXinet	1330	1830
1730	1759	BBCWS-PR	1	The Interview	PI	Sirius141, NPRfm/am, HD+	1330	1830
1730	1759	BBCWS-PR	7	Charlie Gillett's World of Music	MZ	Sirius141, NPRfm/am, HD+	1330	1830
1730	1759	CBC-R1E	1	The Inside Track	SP	CBCLinet, CBCSinet, CBEinet, 1550, CBLAinet, CBMEinet, CBOinet, CBQTinet, CBVEinet, CFFBinet	1330	1330
1730	1759	CBC-R1M	7	The Debaters	LE	CBKinet, 540, CBRinet, 1010, CBXinet, 740, CFYKinet, CHAKinet	1330	1130
1730	1744	CBC-R1S	2-6	Between the Covers	LD	Sirius137	1330	1330
1730	1759	DW	1	A World of Music	MV	DWinet	1330	1930
1730	1559	DW	2	Eurovox	AC	DWinet	1330	1930
1730	1759	DW	3	Hits in Germany	MP	DWinet	1330	1930
1730	1759	DW	4	Arts on the Air	AC	DWinet	1330	1930
1730	1759	DW	5	Cool	CS	DWinet	1330	1930
1730	1759	DW	6	Dialogue	CS	DWinet	1330	1930
1730	1759	DW	7	Insight	NA	DWinet	1330	1930
1730	1759	PR-EXT	2	Talking Jazz	MJ	PR-EXTinet	1330	1930
1730	1759	PR-EXT	3	The Biz	BE	PR-EXTinet	1330	1930
1730	1759	PR-EXT	6	In Touch	LI	PR-EXTinet	1330	1930
1730	1759	PR-EXT	7	Open Air	CS	PR-EXTinet	1330	1930
1730	1759	RNW	1	Reloaded	VA	RNW2inet	1330	1930
1730	1759	RNW	2	Curious Orange	DL	RNW2inet	1330	1930
1730	1759	RNW	3	The State We're In Midweek Edition	CS	RNW2inet	1330	1930
1730	1759	RNW	4	Radio Books	LD	RNW2inet	1330	1930
1730	1759	RNW	5	Earthbeat	EV	RNW2inet	1330	1930
1730	1759	RNW	6	Bridges with Africa	CS	RNW2inet	1330	1930
1730	1749	RNZI	2-5	(see RNZ-NAT)		RNZIinet, 6170w, 7145s	1330	0530
1730	1759	RTE-R1	2-6	Drivetime Sport	SP	RTE-R1inet	1330	1830
1730	1759	WRN-NA	1/7	Glenn Hauser's World of Radio	DX	Sirius140, WRN-NAinet, 9955 (M-F)	1330	1830
1730	1759	WRN-NA	2-6	Radio Netherlands Worldwide features	VA	Sirius140, WRN-NAinet, 9955 (M-F)	1330	1830
1734	1759	ABC-RA	1	Rear Vision	NA	RAinet	1334	0334
1735	1759	PR-EXT	1	Chart Show	MP	PR-EXTinet	1335	1935
1735	1759	PR-EXT	4	Multimedia	DX	PR-EXTinet	1335	1935
1735	1759	PR-EXT	5	High Note	MC	PR-EXTinet	1335	1935
1735	1759	RNZ-NAT	6	Digital Planet	DX	RNZ-NATinet	1335	0535
1745	1759	CBC-R1S	2-6	Outfront	DL	Sirius137	1345	1345
1750	1759	RNZ-NAT	7	Auckland Stories	DL	RNZ-NATinet	1350	0550
1750	1754	RNZI	2-6	New Zealand Coastal Forecast	WX	RNZIinet, 6170w, 7145s	1350	0550
1755	1759	ABC-RN	2-6	Perspective	NC	ABC-RNinet	1355	0355
1755	1759	CRI-RTC	1-7	China Studio	LL	CRI-RTCinet	1355	0155
1755	1759	RNZI	2-6	Pacific Weather Forecast	WX	RNZIinet, 6170w, 7145s	1355	0555
1800	2059	ABC-RA	1-5	Pacific Beat - Morning Edition	NZ	RAinet	1400	0400
1800	1829	ABC-RA	6	Pacific Review	NZ	RAinet	1400	0400
1800	1829	ABC-RA	7	Correspondents Report	NA	RAinet	1400	0400
1800	1859	ABC-RN	1	AWAYE!	DL	ABC-RNinet	1400	0400
1800	1859	ABC-RN	2	The Science Show	ST	ABC-RNinet	1400	0400
1800	1859	ABC-RN	3	Background Briefing	ND	ABC-RNinet	1400	0400
1800	1859	ABC-RN	4	Encounter	CS	ABC-RNinet	1400	0400

UTC Time Start	End	Station/ Network	Day(s)	Program Name	Type	Frequncy/ Platform	EDT	Station Time
1800	1834	ABC-RN	5	MovieTime	AC	ABC-RNinet	1400	0400
1800	1859	ABC-RN	6	Music Deli	MZ	ABC-RNinet	1400	0400
1800	1859	ABC-RN	7	Late Night Live Classic	CA	ABC-RNinet	1400	0400
1800	1814	BBC-R4	1-6	The Archers	LD	BBC-R4inet	1400	1900
1800	1814	BBC-R4	7	From Fact to Fiction	LD	BBC-R4inet	1400	1900
1800	1829	BBCWS-AM	1	World Briefing	NX	XM131	1400	1900
1800	1819	BBCWS-AM	2-6	World Briefing	NX	XM131	1400	1900
1800	1829	BBCWS-AM	7	From Our Own Correspondent	NA	XM131	1400	1900
1800	1829	BBCWS-IE	1	World Briefing	NX	BBCWS-IEinet	1400	1900
1800	1819	BBCWS-IE	2-6	World Briefing	NX	BBCWS-IEinet	1400	1900
1800	1829	BBCWS-IE	7	Documentary feature or series	ND	BBCWS-IEinet	1400	1900
1800	1829	BBCWS-NX	1	World Briefing	NX	BBCWS-NXinet	1400	1900
1800	1819	BBCWS-NX	2-6	World Briefing	NX	BBCWS-NXinet	1400	1900
1800	1829	BBCWS-NX	7	From Our Own Correspondent	NA	BBCWS-NXinet	1400	1900
1800	1829	BBCWS-PR	1	World Briefing	NX	Sirius141, NPRfm/am, HD+	1400	1900
1800	1819	BBCWS-PR	2-6	World Briefing	NX	Sirius141, NPRfm/am, HD+	1400	1900
1800	1829	BBCWS-PR	7	From Our Own Correspondent	NA	Sirius141, NPRfm/am, HD+	1400	1900
1800	1859	CBC-R1A	1	Writers and Company	AC	CBAMinet, CBCTinet, CBDinet, CBHAinet, CBIinet, CBNinet, 640, 6160, CBTinet, CBYinet, CBZFinet, CFGBinet	1400	1500
1800	1829	CBC-R1A	2	Laugh Out Loud	LE	CBAMinet, CBCTinet, CBDinet, CBHAinet, CBIinet, CBNinet, 640, 6160, CBTinet, CBYinet, CBZFinet, CFGBinet	1400	1500
1800	1829	CBC-R1A	3	The Choice	GI	CBAMinet, CBCTinet, CBDinet, CBHAinet, CBIinet, CBNinet, 640, 6160, CBTinet, CBYinet, CBZFinet, CFGBinet	1400	1500
1800	1829	CBC-R1A	4	The Inside Track	SP	CBAMinet, CBCTinet, CBDinet, CBHAinet, CBIinet, CBNinet, 640, 6160, CBTinet, CBYinet, CBZFinet, CFGBinet	1400	1500
1800	1829	CBC-R1A	5	The Age of Persuasion	CS	CBAMinet, CBCTinet, CBDinet, CBHAinet, CBIinet, CBNinet, 640, 6160, CBTinet, CBYinet, CBZFinet, CFGBinet	1400	1500
1800	1829	CBC-R1A	6	Festival of Funny	LE	CBAMinet, CBCTinet, CBDinet, CBHAinet, CBIinet, CBNinet, 640, 6160, CBTinet, CBYinet, CBZFinet, CFGBinet	1400	1500
1800	1859	CBC-R1A	7	The Next Chapter	AC	CBAMinet, CBCTinet, CBDinet, CBHAinet, CBIinet, CBNinet, 640, 6160, CBTinet, CBYinet, CBZFinet, CFGBinet	1400	1500
1800	1829	CBC-R1C	1	WireTap	LE	CBWinet, 990	1400	1300
1800	1959	CBC-R1C	7	Definitely Not the Opera	CS	CBWinet, 990	1400	1300
1800	1859	CBC-R1E	1	Tapestry	CS	CBCLinet, CBCSinet, CBEinet, 1550, CBLAinet, CBMEinet, CBOinet, CBQTinet, CBVEinet, CFFBinet	1400	1400
1800	1859	CBC-R1E	2-6	The Point	CA	CBCLinet, CBCSinet, CBEinet, 1550, CBLAinet, CBMEinet, CBOinet, CBQTinet, CBVEinet, CFFBinet	1400	1400
1800	1859	CBC-R1M	1	The Vinyl Cafe	LE	CBKinet, 540, CBRinet, 1010, CBXinet, 740, CFYKinet, CHAKinet	1400	1200
1800	1959	CBC-R1M	2-6	Local noon-hour program	NZ	CBKinet, 540, CBRinet, 1010, CBXinet, 740, CFYKinet, CHAKinet	1400	1200
1800	1859	CBC-R1M	7	Quirks and Quarks	ST	CBKinet, 540, CBRinet, 1010, CBXinet, 740, CFYKinet, CHAKinet	1400	1200
1800	1829	CBC-R1P	2	White Coat, Black Art	HM	CBCVinet, CBTKinet, CBUinet, 690, 6160, CBYGinet, CFWHinet	1400	1100
1800	1829	CBC-R1P	3	C'est la vie	DL	CBCVinet, CBTKinet, CBUinet, 690, 6160, CBYGinet, CFWHinet	1400	1100
1800	1829	CBC-R1P	4	Spark	DX	CBCVinet, CBTKinet, CBUinet, 690, 6160, CBYGinet, CFWHinet	1400	1100
1800	1829	CBC-R1P	5	Afghanada	LD	CBCVinet, CBTKinet, CBUinet, 690, 6160, CBYGinet, CFWHinet	1400	1100
1800	1859	CBC-R1S	2-6	Rewind	GI	Sirius137	1400	1400
1800	1829	CBC-R1S	7	Spark	DX	Sirius137	1400	1400
1800	1829	CRI-RTC	1-7	News and Reports	NZ	CRI-RTCinet	1400	0200
1800	1805	DW	1-7	News	NX	DWinet	1400	2000
1800	1804	R.PRG	1/7	News	NX	R.PRGinet	1400	2000
1800	1809	R.PRG	2-6	News and Current Affairs	NX	R.PRGinet	1400	2000
1800	1859	RAE	2-6	News, Reports, Features, Tangos	NZ	9690, 15345, RAEinet	1400	1500
1800	1805	RNW	1/7	News	NX	RNW2inet	1400	2000
1800	1829	RNW	2-6	Newsline	NZ	RNW2inet	1400	2000
1800	2059	RNZ-NAT	1-5	Morning Report	NZ	RNZ-NATinet	1400	0600
1800	1859	RNZ-NAT	6/7	Storytime	LD	RNZ-NATinet	1400	0600
1800	1809	RNZI	1-6	World and Pacific News	NX	RNZIinet, 7145s, 11725w	1400	0600
1800	1844	RTE-R1	1	Documentary on One	ND	RTE-R1inet	1400	1900
1800	1859	RTE-R1	2-6	The Dave Fanning Show	AC	RTE-R1inet	1400	1900
1800	1829	RTE-R1	7	Off the Shelf	AC	RTE-R1inet	1400	1900
1800	2229	RTHK-3	1-5	Night Music	MC	RTHK-3inet	1400	0200
1800	2159	RTHK-3	6/7	Night Music	MC	RTHK-3inet	1400	0200
1800	1859	SABC-SAFM	1	Sunday PM	NZ	SABC-SAFMinet	1400	2000
1800	1859	SABC-SAFM	7	Saturday PM	NZ	SABC-SAFMinet	1400	2000
1800	1829	WRN-NA		RTE Ireland: Documentaries	ND	Sirius140, WRN-NAinet, 9955 (M-F)	1400	1900
1800	1829	WRN-NA	2-6	RTE Ireland: Drivetime	NZ	Sirius140, WRN-NAinet, 9955 (M-F)	1400	1900
1800	1859	WRS-SUI	7	News and Music	MP	WRS-SUIinet	1400	2000
1805	1829	DW	1-7	Newslink	NZ	DWinet	1405	2005
1805	1814	R.PRG	1	Mailbox	LI	R.PRGinet	1405	2005
1805	1814	R.PRG	7	Magazine	DL	R.PRGinet	1405	2005
1805	1829	RNW	1	Bridges with Africa	CS	RNW2inet	1405	2005
1805	1859	RNW	7	Network Europe Week	NA	RNW2inet	1405	2005
1810	1819	R.PRG	2	One on One	PI	R.PRGinet	1410	2010
1810	1828	R.PRG	3	Talking Point	DL	R.PRGinet	1410	2010
1810	1828	R.PRG	4	Czechs in History (monthly)	AC	R.PRGinet	1410	2010

UTC Time Start	End	Station/ Network	Day(s)	Program Name	Type	Frequncy/ Platform	EDT	Station Time
1810	1828	R.PRG	4	Czechs Today (monthly)	CS	R.PRGinet	1410	2010
1810	1828	R.PRG	4	Spotlight (fortnightly)	TR	R.PRGinet	1410	2010
1810	1818	R.PRG	5	Panorama	CS	R.PRGinet	1410	2010
1810	1818	R.PRG	6	Business News	BE	R.PRGinet	1410	2010
1810	1814	RNZI	1-6	Sports News	SP	RNZIinet, 7145s, 11725w	1410	0610
1815	1844	BBC-R4	1	Go 4 It!	VA	BBC-R4inet	1415	1915
1815	1844	BBC-R4	2-6	Front Row	AC	BBC-R4inet	1415	1915
1815	1859	BBC-R4	7	Saturday Review	AC	BBC-R4inet	1415	1915
1815	1819	R.PRG	1	Letter from Prague	DL	R.PRGinet	1415	2015
1815	1818	R.PRG	7	Sound Czech	LL	R.PRGinet	1419	2019
1815	1859	RNZI	1	Tagata o te Moana	DL	RNZIinet, 7145s, 11725w	1415	0615
1815	1834	RNZI	2-6	Dateline Pacific	NA	RNZIinet, 7145s, 11725w	1415	0615
1819	1828	R.PRG	1	Czech Books (fortnightly)	AC	R.PRGinet	1419	2019
1819	1828	R.PRG	1	Magic Carpet (monthly)	MZ	R.PRGinet	1419	2019
1819	1828	R.PRG	1	Music Profile (monthly)	MX	R.PRGinet	1419	2019
1819	1828	R.PRG	6	The Arts	AC	R.PRGinet	1419	2019
1819	1828	R.PRG	7	One on One	PI	R.PRGinet	1419	2019
1820	1829	BBCWS-AM	2-6	World Business Report	BE	XM131	1420	1920
1820	1829	BBCWS-IE	2-6	World Business Report	BE	BBCWS-IEinet	1420	1920
1820	1829	BBCWS-NX	2-6	World Business Report	BE	BBCWS-NXinet	1420	1920
1820	1829	BBCWS-PR	2-6	World Business Report	BE	Sirius141, NPRfm/am, HD+	1420	1920
1820	1828	R.PRG	2	Sports News	SP	R.PRGinet	1420	2020
1830	1859	ABC-RA	6	Australian Bite	DL	RAinet	1430	0430
1830	2159	ABC-RA	7	Australia All Over	DL	RAinet	1430	0430
1830	1859	BBCWS-AM	1	Sportsworld Extra	SP	XM131	1430	1930
1830	1859	BBCWS-AM	2-6	Outlook	GZ	XM131	1430	1930
1830	1859	BBCWS-AM	7	Politics UK	GL	XM131	1430	1930
1830	1859	BBCWS-IE	1	Sportsworld Extra	SP	BBCWS-IEinet	1430	1930
1830	1859	BBCWS-IE	2-6	The Strand	AC	BBCWS-IEinet	1430	1930
1830	1839	BBCWS-IE	7	The Instant Guide	NA	BBCWS-IEinet	1430	1930
1830	1859	BBCWS-NX	1	Sportsworld Extra	SP	BBCWS-NXinet	1430	1930
1830	1839	BBCWS-NX	2-6	World Briefing	NX	BBCWS-NXinet	1430	1930
1830	1859	BBCWS-NX	7	Politics UK	GL	BBCWS-NXinet	1430	1930
1830	1859	BBCWS-PR	1	Business Weekly	BE	Sirius141, NPRfm/am, HD+	1430	1930
1830	1859	BBCWS-PR	2	From Our Own Correspondent	NA	Sirius141, NPRfm/am, HD+	1430	1930
1830	1859	BBCWS-PR	3	Heart and Soul	CS	Sirius141, NPRfm/am, HD+	1430	1930
1830	1859	BBCWS-PR	4	The Interview	PI	Sirius141, NPRfm/am, HD+	1430	1930
1830	1859	BBCWS-PR	5	One Planet	EV	Sirius141, NPRfm/am, HD+	1430	1930
1830	1859	BBCWS-PR	6	Global Business	BE	Sirius141, NPRfm/am, HD+	1430	1930
1830	1859	BBCWS-PR	7	Politics UK	GL	Sirius141, NPRfm/am, HD+	1430	1930
1830	1859	CBC-R1C	1	The Inside Track	SP	CBWinet, 990	1430	1330
1830	1859	CBC-R1P	7	The Debaters	LE	CBCVinet, CBTKinet, CBUinet, 690, 6160, CBYGinet, CFWHinet	1430	1130
1830	1859	CBC-R1S	7	White Coat, Black Art	HM	Sirius137	1430	1430
1830	1854	CRI-RTC	1	China Horizons	DL	CRI-RTCinet	1430	0230
1830	1854	CRI-RTC	2	Frontline	GL	CRI-RTCinet	1430	0230
1830	1854	CRI-RTC	3	Biz China	BE	CRI-RTCinet	1430	0230
1830	1854	CRI-RTC	4	In the Spotlight	DL	CRI-RTCinet	1430	0230
1830	1854	CRI-RTC	5	Voices from Other Lands	DL	CRI-RTCinet	1430	0230
1830	1854	CRI-RTC	6	Life in China	DL	CRI-RTCinet	1430	0230
1830	1854	CRI-RTC	7	Listeners' Garden	LI	CRI-RTCinet	1430	0230
1830	1859	DW	1	Insight	NA	DWinet	1430	2030
1830	1859	DW	2	World in Progress	BE	DWinet	1430	2030
1830	1859	DW	3	Spectrum	ST	DWinet	1430	2030
1830	1859	DW	4	Money Talks	BE	DWinet	1430	2030
1830	1859	DW	5	Living Planet	EV	DWinet	1430	2030
1830	1859	DW	6	Inside Europe	NZ	DWinet	1430	2030
1830	1859	DW	7	Dialogue	CS	DWinet	1430	2030
1830	1859	RNW	1/6	Radio Books	LD	RNW2inet	1430	2030
1830	1859	RNW	2	Earthbeat	EV	RNW2inet	1430	2030
1830	1859	RNW	3	Bridges with Africa	CS	RNW2inet	1430	2030
1830	1859	RNW	4/7	Curious Orange	DL	RNW2inet	1430	2030
1830	1859	RNW	5	Network Europe Extra	AC	RNW2inet	1430	2030
1830	1859	RTE-R1	7	The Poetry Programme	LD	RTE-R1inet	1430	1930
1830	1859	WRN-NA	1-7	Radio Prague features	VA	Sirius140, WRN-NAinet, 9955 (M-F)	1430	1930
1835	1859	ABC-RN	5	In Conversation	ST	ABC-RNinet	1435	0435
1835	1839	RNZI	2-6	News about New Zealand	DL	RNZIinet, 7145s, 11725w	1435	0635
1840	1839	BBCWS-IE	7	Over to You	LI	BBCWS-IEinet	1430	1930
1840	1849	BBCWS-NX	2-6	Analysis	NA	BBCWS-NXinet	1440	1940
1845	1859	BBC-R4	1	Literature or drama program	LD	BBC-R4inet	1445	1945
1845	1859	BBC-R4	2-6	Woman's Hour Drama	LD	BBC-R4inet	1445	1945
1845	1859	RTE-R1	1	Short Story	LD	RTE-R1inet	1445	1945
1850	1859	BBCWS-NX	2-6	Sports Roundup	SP	BBCWS-NXinet	1450	1950
1850	1854	RNZI	2-6	Pacific Weather Forecast	WX	RNZIinet, 7145s, 11725w	1450	0650
1855	1859	CRI-RTC	1-7	Chinese Studio	LL	CRI-RTCinet	1430	0230
1900	1929	ABC-RA	6	Asia Review	CA	RAinet	1500	0500

UTC Time Start	End	Station/ Network	Day(s)	Program Name	Type	Frequncy/ Platform	EDT	Station Time
1900	1929	ABC-RN	1	Rear Vision	NA	ABC-RNinet	1500	0500
1900	1929	ABC-RN	2-6	Asia Pacific	NZ	ABC-RNinet	1500	0500
1900	1929	ABC-RN	7	Verbatim	AC	ABC-RNinet	1500	0500
1900	1929	BBC-R4	1	Feedback	LI	BBC-R4inet	1500	2000
1900	1929	BBC-R4	2/5	Documentary feature or series	ND	BBC-R4inet	1500	2000
1900	1939	BBC-R4	3	File on 4	ND	BBC-R4inet	1500	2000
1900	1944	BBC-R4	4	The Moral Maze	CS	BBC-R4inet	1500	2000
1900	1949	BBC-R4	6	Any Questions?	CA	BBC-R4inet	1500	2000
1900	1959	BBC-R4	7	The Archive Hour	GI	BBC-R4inet	1500	2000
1900	1929	BBCWS-AM	1	Heart and Soul	CS	XM131	1500	2000
1900	1929	BBCWS-AM	2/4/6-7	Documentary feature or series	ND	XM131	1500	2000
1900	1929	BBCWS-AM	3	Global Business	BE	XM131	1500	2000
1900	1929	BBCWS-AM	5	Assignment	ND	XM131	1500	2000
1900	1959	BBCWS-IE	1	The Forum	CA	BBCWS-IEinet	1500	2000
1900	1929	BBCWS-IE	2/4/6	Documentary feature or series	ND	BBCWS-IEinet	1500	2000
1900	1929	BBCWS-IE	3	Global Business	BE	BBCWS-IEinet	1500	2000
1900	1929	BBCWS-IE	5	Assignment	ND	BBCWS-IEinet	1500	2000
1900	1959	BBCWS-IE	7	The Strand	AC	BBCWS-IEinet	1500	2000
1900	1929	BBCWS-NX	1/7	World Briefing	NX	BBCWS-NXinet	1500	2000
1900	1919	BBCWS-NX	2-6	World Briefing	NX	BBCWS-NXinet	1500	2000
1900	1929	BBCWS-PR	1/7	World Briefing	NX	Sirius141, NPRfm/am, HD+	1500	2000
1900	1919	BBCWS-PR	2-6	World Briefing	NX	Sirius141, NPRfm/am, HD+	1500	2000
1900	2059	CBC-R1A	2-6	Local afternoon program	NZ	CBAMinet, CBCTinet, CBDinet, CBHAinet, CBIinet, CBNinet, 640, 6160, CBTinet, CBYinet, CBZFinet, CFGBinet	1500	1600
1900	1929	CBC-R1A	7	Spark	DX	CBAMinet, CBCTinet, CBDinet, CBHAinet, CBIinet, CBNinet, 640, 6160, CBTinet, CBYinet, CBZFinet, CFGBinet	1500	1600
1900	1959	CBC-R1C	1	Tapestry	CS	CBWinet, 990	1500	1400
1900	1959	CBC-R1C	2-6	The Point	CA	CBWinet, 990	1500	1400
1900	1959	CBC-R1E	1	Writers and Company	AC	CBCLinet, CBCSinet, CBEinet, 1550, CBLAinet, CBMEinet, CBOinet, CBQTinet, CBVEinet, CFFBinet	1500	1500
1900	1929	CBC-R1E	2	Laugh Out Loud	LE	CBCLinet, CBCSinet, CBEinet, 1550, CBLAinet, CBMEinet, CBOinet, CBQTinet, CBVEinet, CFFBinet	1500	1500
1900	1929	CBC-R1E	3	The Choice	GI	CBCLinet, CBCSinet, CBEinet, 1550, CBLAinet, CBMEinet, CBOinet, CBQTinet, CBVEinet, CFFBinet	1500	1500
1900	1929	CBC-R1E	4	The Inside Track	SP	CBCLinet, CBCSinet, CBEinet, 1550, CBLAinet, CBMEinet, CBOinet, CBQTinet, CBVEinet, CFFBinet	1500	1500
1900	1929	CBC-R1E	5	The Age of Persuasion	DX	CBCLinet, CBCSinet, CBEinet, 1550, CBLAinet, CBMEinet, CBOinet, CBQTinet, CBVEinet, CFFBinet	1500	1500
1900	1929	CBC-R1E	6	Festival of Funny	LE	CBCLinet, CBCSinet, CBEinet, 1550, CBLAinet, CBMEinet, CBOinet, CBQTinet, CBVEinet, CFFBinet	1500	1500
1900	1959	CBC-R1E	7	The Next Chapter	AC	CBCLinet, CBCSinet, CBEinet, 1550, CBLAinet, CBMEinet, CBOinet, CBQTinet, CBVEinet, CFFBinet	1500	1500
1900	1929	CBC-R1M	1	WireTap	LE	CBKinet, 540, CBRinet, 1010, CBXinet, 740, CFYKinet, CHAKinet	1500	1300
1900	2059	CBC-R1M	7	Definitely Not the Opera	CS	CBKinet, 540, CBRinet, 1010, CBXinet, 740, CFYKinet, CHAKinet	1500	1300
1900	1959	CBC-R1P	1	The Vinyl Cafe	LE	CBCVinet, CBTKinet, CBUinet, 690, 6160, CBYGinet, CFWHinet	1500	1200
1900	2059	CBC-R1P	2-6	Local noon-hour program	NZ	CBCVinet, CBTKinet, CBUinet, 690, 6160, CBYGinet, CFWHinet	1500	1200
1900	1959	CBC-R1P	7	Quirks and Quarks	ST	CBCVinet, CBTKinet, CBUinet, 690, 6160, CBYGinet, CFWHinet	1500	1200
1900	1959	CBC-R1S	1	The Vinyl Cafe	LE	Sirius137	1500	1500
1900	2029	CBC-R1S	2-6	Q	AC	Sirius137	1500	1500
1900	1959	CBC-R1S	7	Quirks and Quarks	ST	Sirius137	1500	1500
1900	1929	CRI-RTC	1-7	News and Reports	NZ	CRI-RTCinet	1500	0300
1900	1905	DW	1-7	News	NX	DWinet	1500	2100
1900	1905	RNW	1/7	News	NX	RNW2inet	1500	2100
1900	1929	RNW	2-6	Newsline	NZ	RNW2inet	1500	2100
1900	1959	RNZ-NAT	6	Country Life	DL	RNZ-NATinet	1500	0700
1900	1934	RNZ-NAT	7	Hymns for Sunday Morning	CS	RNZ-NATinet	1500	0700
1900	1909	RNZI	1-6	World and Pacific News	NX	RNZIinet, 9615s, 11725w	1500	0700
1900	1959	RTE-R1	1	Documentary series	GD	RTE-R1inet	1500	2000
1900	1959	RTE-R1	2-6	The Arts Show	AC	RTE-R1inet	1500	2000
1900	1959	RTE-R1	7	South Wind Blows	MF	RTE-R1inet	1500	2000
1900	2159	SABC-SAFM	1	Classical Sunday	MC	SABC-SAFMinet	1500	2100
1900	1959	SABC-SAFM	2	Law Report	GL	SABC-SAFMinet	1500	2100
1900	1959	SABC-SAFM	3	Health Hour	HM	SABC-SAFMinet	1500	2100
1900	1959	SABC-SAFM	4	Time to Travel	TR	SABC-SAFMinet	1500	2100
1900	1959	SABC-SAFM	5	Earth Hour	ST	SABC-SAFMinet	1500	2100
1900	1959	SABC-SAFM	6	Art Matters	AC	SABC-SAFMinet	1500	2100
1900	2159	SABC-SAFM	7	Best of Jazz	MJ	SABC-SAFMinet	1500	2100
1900	1929	WRN-NA	1-7	Radio Sweden features	VA	Sirius140, WRN-NAinet, 9955 (M-F)	1500	2000
1900	1959	WRS-SUI	1	News and Music	MP	WRS-SUIinet	1500	2100
1900	1959	WRS-SUI	7	World Drama	LD	WRS-SUIinet	1500	2100
1905	1929	DW	1-7	Newslink	NZ	DWinet	1505	2105
1905	1929	RNW	1	Network Europe Extra	AC	RNW2inet	1505	2105
1905	1959	RNW	7	The State We're In	CS	RNW2inet	1505	2105
1910	1929	RNZI	1-5	(see RNZ-NAT)		RNZIinet, 9615s, 11725w	1510	0710

UTC Time Start	End	Station/ Network	Day(s)	Program Name	Type	Frequncy/ Platform	EDT	Station Time
1910	1959	RNZI	6	Tagata o te Moana	DL	RNZIinet, 9615s, 11725w	1510	0710
1920	1929	BBCWS-NX	2-6	World Business Report	BE	BBCWS-NXinet	1520	2020
1920	1929	BBCWS-PR	2-6	World Business Report	BE	Sirius141, NPRfm/am, HD+	1520	2020
1930	1959	ABC-RA	6	Rural Reporter	DL	RAinet	1530	0530
1930	1959	ABC-RN	1/2/6	BBC Comedy	LE	ABC-RNinet	1530	0530
1930	1959	ABC-RN	3	My Music	LE	ABC-RNinet	1530	0530
1930	1959	ABC-RN	4	My Word	LE	ABC-RNinet	1530	0530
1930	1959	ABC-RN	5	The Goons	LE	ABC-RNinet	1530	0530
1930	1959	ABC-RN	7	Radio National Quiz	QG	ABC-RNinet	1530	0530
1930	1959	BBC-R4	1	Last Word	CS	BBC-R4inet	1530	2030
1930	1959	BBC-R4	2	The Learning Curve	CS	BBC-R4inet	1530	2030
1930	1959	BBC-R4	5	In Business	BE	BBC-R4inet	1530	2030
1930	1959	BBCWS-AM	1	One Planet	EV	XM131	1530	2030
1930	1959	BBCWS-AM	2-6	The Strand	AC	XM131	1530	2030
1930	1959	BBCWS-AM	7	Business Weekly	BE	XM131	1530	2030
1930	1959	BBCWS-IE	2	Health Check	HM	BBCWS-IEinet	1530	2030
1930	1959	BBCWS-IE	3	Digital Planet	DX	BBCWS-IEinet	1530	2030
1930	1959	BBCWS-IE	4	Discovery	ST	BBCWS-IEinet	1530	2030
1930	1959	BBCWS-IE	5	One Planet	EV	BBCWS-IEinet	1530	2030
1930	1959	BBCWS-IE	6	Science in Action	ST	BBCWS-IEinet	1530	2030
1930	1959	BBCWS-NX	1	Reporting Religion	CS	BBCWS-NXinet	1530	2030
1930	1939	BBCWS-NX	2-6	World Briefing	NX	BBCWS-NXinet	1530	2030
1930	1959	BBCWS-NX	7	The Interview	PI	BBCWS-NXinet	1530	2030
1930	1959	BBCWS-PR	1	Reporting Religion	CS	Sirius141, NPRfm/am, HD+	1530	2030
1930	1939	BBCWS-PR	2-6	World Briefing	NX	Sirius141, NPRfm/am, HD+	1530	2030
1930	1959	BBCWS-PR	7	The Interview	PI	Sirius141, NPRfm/am, HD+	1530	2030
1930	1959	CBC-R1A	7	White Coat, Black Art	HM	CBAMinet, CBCTinet, CBDinet, CBHAinet, CBlinet, CBNinet, 640, 6160, CBTinet, CBYinet, CBZFinet, CFGBinet	1530	1630
1930	1959	CBC-R1M	1	The Inside Track	SP	CBKinet, 540, CBRinet, 1010, CBXinet, 740, CFYKinet, CHAKinet	1530	1330
1930	1959	CRI-RTC	1	China Horizons	DL	CRI-RTCinet	1530	0330
1930	1959	CRI-RTC	2	Frontline	GL	CRI-RTCinet	1530	0330
1930	1959	CRI-RTC	3	Biz China	BE	CRI-RTCinet	1530	0330
1930	1959	CRI-RTC	4	In the Spotlight	DL	CRI-RTCinet	1530	0330
1930	1959	CRI-RTC	5	Voices from Other Lands	DL	CRI-RTCinet	1530	0330
1930	1959	CRI-RTC	6	Life in China	DL	CRI-RTCinet	1530	0330
1930	1959	CRI-RTC	7	Listeners' Garden	LI	CRI-RTCinet	1530	0330
1930	1944	DW	1/7	Sports Report	SP	DWinet	1530	2130
1930	1959	DW	2	Eurovox	AC	DWinet	1530	2130
1930	1959	DW	3	Hits in Germany	MP	DWinet	1530	2130
1930	1959	DW	4	Arts on the Air	AC	DWinet	1530	2130
1930	1959	DW	5	Cool	CS	DWinet	1530	2130
1930	1959	DW	6	Dialogue	CS	DWinet	1530	2130
1930	2029	IRIB	1-7	Koran Reading, News and features	VA	IRIBinet	1530	2300
1930	1954	PR-EXT	1	Network Europe Extra	AC	PR-EXTinet	1530	2130
1930	1949	PR-EXT	2-6	News from Poland	NX	PR-EXTinet	1530	2130
1930	1949	PR-EXT	7	Europe East	NZ	PR-EXTinet	1530	2130
1930	1959	RNW	1	Reloaded	VA	RNW2inet	1530	2130
1930	1959	RNW	2	Curious Orange	DL	RNW2inet	1530	2130
1930	1959	RNW	3	The State We're In Midweek Edition	CS	RNW2inet	1530	2130
1930	1959	RNW	4	Radio Books	LD	RNW2inet	1530	2130
1930	1959	RNW	5	Earthbeat	EV	RNW2inet	1530	2130
1930	1959	RNW	6	Bridges with Africa	CS	RNW2inet	1530	2130
1930	1934	RNZI	1-5	New Zealand News Headlines	NX	RNZIinet, 9615s, 11725w	1530	0730
1930	1959	WRN-NA	1	Radio New Zealand Int.: Dateline Pacific	NZ	Sirius140, WRN-NAinet, 9955 (M-F)	1530	2030
1930	1959	WRN-NA	2-7	Radio Australia features	VA	Sirius140, WRN-NAinet, 9955 (M-F)	1530	2030
1930	1959	WRS-SUI	2-6	Music	MP	WRS-SUIinet	1530	2130
1935	1959	RNZ-NAT	7	Weekend Worldwatch	NA	RNZ-NATinet	1535	0735
1935	1959	RNZI	1-5	Pacific Business Report	BE	RNZIinet, 9615s, 11725w	1535	0735
1935	1959	RNZI	7	World Watch	NA	RNZIinet, 9615s, 11725w	1535	0735
1940	1959	BBC-R4	3	In Touch	CS	BBC-R4inet	1540	2040
1940	1949	BBCWS-NX	2-6	Analysis	NA	BBCWS-NXinet	1540	2040
1940	1959	BBCWS-PR	2-6	Business Daily	BE	Sirius141, NPRfm/am, HD+	1540	2040
1945	1959	BBC-R4	4	Documentary feature or series	ND	BBC-R4inet	1545	2045
1945	1959	DW	1	Inspired Minds	AC	DWinet	1545	2145
1945	1959	DW	7	Radio D	LL	DWinet	1545	2145
1950	1959	BBC-R4	6	A Point of View	NC	BBC-R4inet	1550	2050
1950	1959	BBCWS-NX	2-6	Sports Roundup	SP	BBCWS-NXinet	1540	2040
1950	1959	PR-EXT	2	Around Poland	TR	PR-EXTinet	1550	2150
1950	1959	PR-EXT	3	Letter from Poland	DL	PR-EXTinet	1550	2150
1950	2004	PR-EXT	4	Day in the Life	PI	PR-EXTinet	1550	2150
1950	1954	PR-EXT	5	Comment	NC	PR-EXTinet	1550	2150
1950	1959	PR-EXT	6	Business Week	BE	PR-EXTinet	1550	2150
1950	1959	PR-EXT	7	A Look at the Weeklies	PR	PR-EXTinet	1550	2150
1955	2004	PR-EXT	1	The Kids	CS	PR-EXTinet	1555	2155
1955	2004	PR-EXT	5	Focus	AC	PR-EXTinet	1555	2155
2000	2029	ABC-RA	6	Pacific Review	NZ	RAinet	1600	0600

UTC Time Start	End	Station/ Network	Day(s)	Program Name	Type	Frequncy/ Platform	EDT	Station Time
2000	2059	ABC-RN	1-5	Radio National Breakfast	NZ	ABC-RNinet	1600	0600
2000	2059	ABC-RN	6	Country Breakfast	DL	ABC-RNinet	1600	0600
2000	2059	ABC-RN	7	Life and Times	AC	ABC-RNinet	1600	0600
2000	2025	BBC-R4	1	Money Box	BE	BBC-R4inet	1600	2100
2000	2029	BBC-R4	2/3	Science or health series or feature	ST	BBC-R4inet	1600	2100
2000	2029	BBC-R4	4	World on the Move	ST	BBC-R4inet	1600	2100
2000	2029	BBC-R4	5	Leading Edge	ST	BBC-R4inet	1600	2100
2000	2057	BBC-R4	6	Documentary feature or series	ND	BBC-R4inet	1600	2100
2000	2059	BBC-R4	7	The Classic Serial	LD	BBC-R4inet	1600	2100
2000	2059	BBCWS-AM	1-7	Newshour	NZ	XM131	1600	2100
2000	2059	BBCWS-IE	1-7	Newshour	NZ	BBCWS-IEinet	1600	2100
2000	2059	BBCWS-NX	1-7	Newshour	NZ	BBCWS-NXinet	1600	2100
2000	2059	BBCWS-PR	1-7	Newshour	NZ	Sirius141, NPRfm/am, HD+	1600	2100
2000	2159	CBC-R1A	1	Cross Country Checkup	DL	CBAMinet, CBCTinet, CBDinet, CBHAinet, CBIinet, CBNinet, 640, 6160, CBTinet, CBYinet, CBZFinet, CFGBinet	1600	1700
2000	2059	CBC-R1A	7	Regional performance	GI	CBAMinet, CBCTinet, CBDinet, CBHAinet, CBIinet, CBNinet, 640, 6160, CBTinet, CBYinet, CBZFinet, CFGBinet	1600	1700
2000	2159	CBC-R1C	1	Cross Country Checkup	DL	CBWinet, 990	1600	1500
2000	2029	CBC-R1C	2	Laugh Out Loud	LE	CBWinet, 990	1600	1500
2000	2029	CBC-R1C	3	The Choice	GI	CBWinet, 990	1600	1500
2000	2029	CBC-R1C	4	The Inside Track	SP	CBWinet, 990	1600	1500
2000	2029	CBC-R1C	5	The Age of Persuasion	DX	CBWinet, 990	1600	1500
2000	2029	CBC-R1C	6	Festival of Funny	LE	CBWinet, 990	1600	1500
2000	2059	CBC-R1C	7	The Next Chapter	AC	CBWinet, 990	1600	1500
2000	2159	CBC-R1E	1	Cross Country Checkup	DL	CBCLinet, CBCSinet, CBEinet, 1550, CBLAinet, CBMEinet, CBOinet, CBQTinet, CBVEinet, CFFBinet	1600	1600
2000	2159	CBC-R1E	2-6	Local afternoon program	NZ	CBCLinet, CBCSinet, CBEinet, 1550, CBLAinet, CBMEinet, CBOinet, CBQTinet, CBVEinet, CFFBinet	1600	1600
2000	2029	CBC-R1E	7	Spark	DX	CBCLinet, CBCSinet, CBEinet, 1550, CBLAinet, CBMEinet, CBOinet, CBQTinet, CBVEinet, CFFBinet	1600	1600
2000	2159	CBC-R1M	1	Cross Country Checkup	DL	CBKinet, 540, CBRinet, 1010, CBXinet, 740, CFYKinet, CHAKinet	1600	1400
2000	2059	CBC-R1M	2-6	The Point	CA	CBKinet, 540, CBRinet, 1010, CBXinet, 740, CFYKinet, CHAKinet	1600	1400
2000	2159	CBC-R1P	1	Cross Country Checkup	DL	CBCVinet, CBTKinet, CBUinet, 690, 6160, CBYGinet, CFWHinet	1600	1300
2000	2159	CBC-R1P	7	Definitely Not the Opera	CS	CBCVinet, CBTKinet, CBUinet, 690, 6160, CBYGinet, CFWHinet	1600	1300
2000	2159	CBC-R1S	1	Cross Country Checkup	DL	Sirius137	1600	1600
2000	2159	CBC-R1S	7	Vinyl Tap	MP	Sirius137	1600	1600
2000	2029	CRI-RTC	1-7	News and Reports	NZ	CRI-RTCinet	1600	0400
2000	2029	CRI-WASH	1/7	People in the Know	GL	CRI-WASHinet	1600	0400
2000	2029	CRI-WASH	2	China Horizons	DL	CRI-WASHinet	1600	0400
2000	2029	CRI-WASH	3	Frontline	GL	CRI-WASHinet	1600	0400
2000	2029	CRI-WASH	4	Biz China	BE	CRI-WASHinet	1600	0400
2000	2029	CRI-WASH	5	In the Spotlight	DL	CRI-WASHinet	1600	0400
2000	2029	CRI-WASH	6	Voices from Other Lands	DL	CRI-WASHinet	1600	0400
2000	2005	DW	1-7	News	NX	DWinet	1600	2200
2000	2029	PR-EXT	2	Talking Jazz	MJ	PR-EXTinet	1600	2200
2000	2029	PR-EXT	3	The Biz	BE	PR-EXTinet	1600	2200
2000	2029	PR-EXT	6	In Touch	LI	PR-EXTinet	1600	2200
2000	2029	PR-EXT	7	Open Air	CS	PR-EXTinet	1600	2200
2000	2029	RNW	1	Bridges with Africa	CS	RNW2inet	1600	2200
2000	2029	RNW	2-6	Network Europe	NZ	RNW2inet	1600	2200
2000	2059	RNW	7	Earthbeat	EV	RNW2inet	1600	2200
2000	2359	RNZ-NAT	6	Saturday Morning with Kim Hill	VA	RNZ-NATinet	1600	0800
2000	2359	RNZ-NAT	7	Sunday Morning with Chris Laidlaw	GZ	RNZ-NATinet	1600	0800
2000	2009	RNZI	1-6	World and Pacific News	NX	RNZIinet, 17675, 15720	1600	0800
2000	2059	RNZI	7	Sportsworld	SP	RNZIinet, 17675, 15720	1610	0810
2000	2059	RTE-R1	1	O'Brien on Song	MV	RTE-R1inet	1600	2100
2000	2049	RTE-R1	2-6	The Radio 1 Music Collection	MV	RTE-R1inet	1600	2100
2000	2059	RTE-R1	7	Ceili House	MF	RTE-R1inet	1600	2100
2000	2014	SABC-CHAF	1/7	News	NX	CHAFStudio1inet	1600	2200
2000	2014	SABC-CHAF	2-6	180 Degrees	NX	CHAFStudio1inet	1600	2200
2000	2159	SABC-SAFM	2-6	NightTime Music	MV	SABC-SAFMinet	1600	2200
2000	2029	WRN-NA	1-7	Polish Radio External Service features	VA	Sirius140, WRN-NAinet, 9955 (M-F)	1600	2100
2000	2059	WRS-SUI	1-7	BBC Newshour	NZ	WRS-SUIinet	1600	2200
2005	2029	DW	1-7	Newslink	NZ	DWinet	1605	2205
2005	2029	PR-EXT	1	Chart Show	MP	PR-EXTinet	1605	2205
2005	2029	PR-EXT	4	Multimedia	DX	PR-EXTinet	1605	2205
2005	2029	PR-EXT	5	High Note	MC	PR-EXTinet	1605	2205
2010	2014	RNZI	1-6	Sports News	SP	RNZIinet, 17675, 15720	1610	0810
2012	2039	RNZ-NAT	7	Insight	NA	RNZ-NATinet	1612	0812
2015	2034	RNZI	1-6	Dateline Pacific	NA	RNZIinet, 17675, 15720	1615	0815
2015	2059	SABC-CHAF	1-7	Talk or Debate	GI	CHAFStudio1inet	1615	2215
2026	2029	BBC-R4	1	The Radio 4 Appeal	DL	BBC-R4inet	1626	2126
2030	2059	ABC-RA	6	Australian Country Style	MW	RAinet	1630	0630
2030	2057	BBC-R4	1	In Business	BE	BBC-R4inet	1630	2130
2030	2057	BBC-R4	2	Start the Week	CA	BBC-R4inet	1630	2130

UTC Time Start	End	Station/ Network	Day(s)	Program Name	Type	Frequency/ Platform	EDT	Station Time
2030	2057	BBC-R4	3	General feature or documentary series	GI	BBC-R4inet	1630	2130
2030	2057	BBC-R4	4	Midweek	CA	BBC-R4inet	1630	2130
2030	2057	BBC-R4	5	In Our Time	ED	BBC-R4inet	1630	2130
2030	2059	CBC-R1E	7	White Coat, Black Art	HM	CBCLinet, CBCSinet, CBEinet, 1550, CBLAinet, CBMEinet, CBOinet, CBQTinet, CBVEinet, CFFBinet	1630	1630
2030	2059	CBC-R1S	2	White Coat, Black Art	HM	Sirius137	1630	1630
2030	2059	CBC-R1S	3	C'est la vie	DL	Sirius137	1630	1630
2030	2059	CBC-R1S	4	Spark	DX	Sirius137	1630	1630
2030	2059	CBC-R1S	5	Laugh Out Loud	LE	Sirius137	1630	1630
2030	2059	CBC-R1S	6	WireTap	LE	Sirius137	1630	1630
2030	2054	CRI-RTC	1	China Horizons	DL	CRI-RTCinet	1630	0430
2030	2054	CRI-RTC	2	Frontline	GL	CRI-RTCinet	1630	0430
2030	2054	CRI-RTC	3	Biz China	BE	CRI-RTCinet	1630	0430
2030	2054	CRI-RTC	4	In the Spotlight	DL	CRI-RTCinet	1630	0430
2030	2054	CRI-RTC	5	Voices from Other Lands	DL	CRI-RTCinet	1630	0430
2030	2054	CRI-RTC	6	Life in China	DL	CRI-RTCinet	1630	0430
2030	2054	CRI-RTC	7	Listeners' Garden	LI	CRI-RTCinet	1630	0430
2030	2039	CRI-WASH	1-7	CRI Roundup	NX	CRI-WASHinet	1630	0430
2030	2059	DW	1	Cool	CS	DWinet	1630	2230
2030	2059	DW	2	World in Progress	BE	DWinet	1630	2230
2030	2059	DW	3	Spectrum	ST	DWinet	1630	2230
2030	2059	DW	4	Money Talks	BE	DWinet	1630	2230
2030	2059	DW	5	Living Planet	EV	DWinet	1630	2230
2030	2059	DW	6	Inside Europe	NZ	DWinet	1630	2230
2030	2059	DW	7	Dialogue	CS	DWinet	1630	2230
2030	2039	KBS-WR	1-7	News	NX	KBS-WR2inet	1630	0530
2030	2059	RNW	1/6	Radio Books	LD	RNW2inet	1630	2230
2030	2059	RNW	2	Earthbeat	EV	RNW2inet	1630	2230
2030	2059	RNW	3	Bridges with Africa	CS	RNW2inet	1630	2230
2030	2059	RNW	4/7	Curious Orange	DL	RNW2inet	1630	2230
2030	2059	RNW	5	Network Europe Extra	AC	RNW2inet	1630	2230
2030	2059	WRN-NA	1-7	KBS World Radio features	VA	Sirius140, WRN-NAinet, 9955 (M-F)	1630	2130
2035	2039	RNZI	1-5	News about New Zealand	NX	RNZIinet, 17675, 15720	1635	0835
2035	2059	RNZI	6	Mailbox (fortnightly)	DX	RNZIinet, 17675, 15720	1635	0835
2035	2059	RNZI	6	Spectrum (fortnightly)	DL	RNZIinet, 17675, 15720	1635	0835
2040	2059	CRI-WASH	1-7	Listener's Garden	GZ	CRI-WASHinet	1640	0440
2040	2059	KBS-WR	1	Korean Pop Interactive	MP	KBS-WR2inet	1640	0540
2040	2044	KBS-WR	2-6	News Commentary	NC	KBS-WR2inet	1640	0540
2040	2059	KBS-WR	7	Worldwide Friendship	LI	KBS-WR2inet	1640	0540
2040	2049	RNZI	1	Focus on Politics	GL	RNZIinet, 17675, 15720	1640	0840
2040	2049	RNZI	3-5	RNZI Feature	GI	RNZIinet, 17675, 15720	1640	0840
2045	2059	KBS-WR	2	Faces of Korea	CS	KBS-WR2inet	1645	0545
2045	2059	KBS-WR	3	Business Watch	BE	KBS-WR2inet	1645	0545
2045	2059	KBS-WR	4	Culture on the Move	AC	KBS-WR2inet	1645	0545
2045	2059	KBS-WR	5	Korea Today and Tomorrow	DL	KBS-WR2inet	1645	0545
2045	2059	KBS-WR	6	Seoul Report	DL	KBS-WR2inet	1645	0545
2050	2054	RNZI	1-5	New Zealand Newspaper Headlines	PR	RNZIinet, 17675, 15720	1650	0850
2050	2059	RTE-R1	2-6	Nuacht (in Irish Gaelic)	NX	RTE-R1inet	1650	2150
2055	2059	CRI-RTC	1-7	Chinese Studio	LL	CRI-RTCinet	1630	0430
2055	2059	RNZI	1-5	Pacific Business Report	BE	RNZIinet, 17675, 15720	1655	0855
2058	2059	BBC-R4	1-6	Weather	ED	BBC-R4inet	1658	2158
2100	2129	ABC-RA	1-5	AM	NZ	RAinet	1700	0700
2100	2109	ABC-RA	6	Correspondent's Notebook	NA	RAinet	1700	0700
2100	2129	ABC-RN	1-5	AM	NZ	ABC-RNinet	1700	0700
2100	2129	ABC-RN	6	Saturday AM	NZ	ABC-RNinet	1700	0700
2100	2159	ABC-RN	7	Encounter	CS	ABC-RNinet	1700	0700
2100	2159	BBC-R4	1	The Westminster Hour	GL	BBC-R4inet	1700	2200
2100	2144	BBC-R4	2-6	The World Tonight	NZ	BBC-R4inet	1700	2200
2100	2114	BBC-R4		News and Weather	NX	BBC-R4inet	1700	2200
2100	2129	BBCWS-AM	1/7	Documentary feature or series	ND	XM131	1700	2200
2100	2119	BBCWS-AM	2-6	World Briefing	NX	XM131	1700	2200
2100	2119	BBCWS-IE	1-6	World Briefing	NX	BBCWS-IEinet	1700	2200
2100	2129	BBCWS-IE	7	Documentary feature or series	ND	BBCWS-IEinet	1700	2200
2100	2119	BBCWS-NX	1-6	World Briefing	NX	BBCWS-NXinet	1700	2200
2100	2129	BBCWS-NX	7	Assignment	ND	BBCWS-NXinet	1700	2200
2100	2159	BBCWS-PR	1	The Forum	CA	Sirius141, NPRfm/am, HD+	1700	2200
2100	2119	BBCWS-PR	2-6	World Briefing	NX	Sirius141, NPRfm/am, HD+	1700	2200
2100	2129	BBCWS-PR	7	Assignment	ND	Sirius141, NPRfm/am, HD+	1700	2200
2100	2129	CBC-R1A	2-6	The World At Six	NX	CBAMinet, CBCTinet, CBDinet, CBHAinet, CBIinet, CBNinet, 640, 6160, CBTinet, CBYinet, CBZFinet, CFGBinet	1700	1800
2100	2159	CBC-R1A	7	A Propos	MF	CBAMinet, CBCTinet, CBDinet, CBHAinet, CBIinet, CBNinet, 640, 6160, CBTinet, CBYinet, CBZFinet, CFGBinet	1700	1800
2100	2259	CBC-R1C	2-6	Local afternoon program	NZ	CBWinet, 990	1700	1600

UTC Time Start	End	Station/ Network	Day(s)	Program Name	Type	Frequncy/ Platform	EDT	Station Time
2100	2129	CBC-R1C	7	Spark	DX	CBWinet, 990	1700	1600
2100	2159	CBC-R1E	7	Regional performance	GI	CBCLinet, CBCSinet, CBEinet, 1550, CBLAinet, CBMEinet, CBOinet, CBQTinet, CBVEinet, CFFBinet	1700	1700
2100	2129	CBC-R1M	2	Laugh Out Loud	LE	CBKinet, 540, CBRinet, 1010, CBXinet, 740, CFYKinet, CHAKinet	1700	1500
2100	2129	CBC-R1M	3	The Choice	GI	CBKinet, 540, CBRinet, 1010, CBXinet, 740, CFYKinet, CHAKinet	1700	1500
2100	2129	CBC-R1M	4	The Inside Track	SP	CBKinet, 540, CBRinet, 1010, CBXinet, 740, CFYKinet, CHAKinet	1700	1500
2100	2129	CBC-R1M	5	The Age of Persuasion	DX	CBKinet, 540, CBRinet, 1010, CBXinet, 740, CFYKinet, CHAKinet	1700	1500
2100	2129	CBC-R1M	6	Festival of Funny	LE	CBKinet, 540, CBRinet, 1010, CBXinet, 740, CFYKinet, CHAKinet	1700	1500
2100	2159	CBC-R1M	7	The Next Chapter	AC	CBKinet, 540, CBRinet, 1010, CBXinet, 740, CFYKinet, CHAKinet	1700	1500
2100	2159	CBC-R1P	2-6	The Point	CA	CBCVinet, CBTKinet, CBUinet, 690, 6160, CBYGinet, CFWHinet	1700	1400
2100	2129	CBC-R1S	2-6	The World At Six	NX	Sirius137	1700	1700
2100	2129	CRI-RTC	1-7	News and Reports	NZ	CRI-RTCinet	1700	0500
2100	2229	CRI-WASH	1-7	China Drive	NZ	CRI-WASHinet	1700	0500
2100	2105	DW	1-7	News	NX	DWinet	1700	2300
2100	2104	R.PRG	1/7	News	NX	R.PRGinet, 5930	1700	2300
2100	2109	R.PRG	2-6	News and Current Affairs	NX	R.PRGinet, 5930	1700	2300
2100	2129	RNW	1	Network Europe Extra	AC	RNW2inet	1700	2300
2100	2129	RNW	2-6	Newsline	NZ	RNW2inet	1700	2300
2100	2129	RNW	7	The State We're In	CS	RNW2inet	1700	2300
2100	0059	RNZ-NAT	1-5	Nine to Noon with Kathryn Ryan	CA	RNZ-NATinet	1700	0900
2100	2109	RNZI	1-6	World and Pacific News	NX	RNZIinet, 17675, 15720	1700	0900
2100	2359	RNZI	7	(see RNZ-NAT)		RNZIinet, 15720, 17675	1700	0900
2100	2159	RTE-R1	1	The Late Session	MF	RTE-R1inet	1700	2200
2100	2159	RTE-R1	2-5	Feature/Arts/Documentary series	DC	RTE-R1inet	1700	2200
2100	2129	RTE-R1	6	Farm Week	BE	RTE-R1inet	1700	2200
2100	2159	RTE-R1	7	Failte Isteach	MF	RTE-R1inet	1700	2200
2100	2159	SABC-CHAF	1	African Music (News on the hour)	MF	CHAFStudio1inet	1700	2300
2100	2114	SABC-CHAF	1	News	NX	CHAFStudio1inet	1700	2300
2100	2114	SABC-CHAF	2-6	180 Degrees	NX	CHAFStudio1inet	1700	2300
2100	2129	WRN-NA	1	RTE Ireland feature	VA	Sirius140, WRN-NAinet	1700	2200
2100	2129	WRN-NA	2-6	RTE Ireland: Drivetime	NZ	Sirius140, WRN-NAinet	1700	2200
2100	2129	WRN-NA	2-6	RTE Ireland: Sport	SP	Sirius140, WRN-NAinet	1700	2200
2100	2159	WRS-SUI	1-7	Music	MP	WRS-SUIinet	1700	2300
2105	2129	DW	1-7	Newslink	NZ	DWinet	1705	2305
2105	2114	R.PRG	1	Mailbox	LI	R.PRGinet, 5930	1705	2305
2105	2114	R.PRG	7	Magazine	DL	R.PRGinet, 5930	1705	2305
2110	2129	ABC-RA	6	Saturday AM	NZ	RAinet	1710	0710
2110	2119	R.PRG	2	One on One	PI	R.PRGinet, 5930	1710	2310
2110	2128	R.PRG	3	Talking Point	DL	R.PRGinet, 5930	1710	2310
2110	1828	R.PRG	4	Czechs in History (monthly)	AC	R.PRGinet, 5930	1710	2310
2110	1828	R.PRG	4	Czechs Today (monthly)	CS	R.PRGinet, 5930	1710	2310
2110	1828	R.PRG	4	Spotlight (fortnightly)	TR	R.PRGinet, 5930	1710	2310
2110	1818	R.PRG	5	Panorama	CS	R.PRGinet, 5930	1710	2310
2110	1818	R.PRG	6	Business News	BE	R.PRGinet, 5930	1710	2310
2110	2130	RNZ-NAT	7	Mediawatch	DX	RNZ-NATinet	1710	0910
2110	2114	RNZI	1-5	Sports News	SP	RNZIinet, 15720	1710	0910
2110	2359	RNZI	6	(see RNZ-NAT)		RNZIinet, 15720, 17675	1710	0910
2115	2159	BBC-R4	7	The Moral Maze	CS	BBC-R4inet	1715	2215
2115	1819	R.PRG	1	Letter from Prague	DL	R.PRGinet, 5930	1715	2315
2115	2118	R.PRG	7	Sound Czech	LL	R.PRGinet, 5930	1719	2319
2115	2134	RNZI	3	Tradewinds	BE	RNZIinet, 17675	1715	0915
2115	2134	RNZI	4	World in Sport	SP	RNZIinet, 17675	1715	0915
2115	2134	RNZI	5	Pacific Correspondent	NA	RNZIinet, 17675	1715	0915
2115	2159	SABC-CHAF	2-7	Talk or Debate	GI	CHAFStudio1inet	1715	2315
2119	2128	R.PRG	1	Czech Books (fortnightly)	AC	R.PRGinet, 5930	1719	2319
2119	2128	R.PRG	1	Magic Carpet (monthly)	MZ	R.PRGinet, 5930	1719	2319
2119	2128	R.PRG	1	Music Profile (monthly)	MX	R.PRGinet, 5930	1719	2319
2119	2128	R.PRG	6	The Arts	AC	R.PRGinet, 5930	1719	2319
2119	2128	R.PRG	7	One on One	PI	R.PRGinet, 5930	1719	2319
2120	2129	BBCWS-AM	2-6	World Business Report	BE	XM131	1720	2220
2120	2129	BBCWS-IE	1	Sports Roundup	SP	BBCWS-IEinet	1720	2220
2120	2129	BBCWS-IE	2-6	Analysis	NA	BBCWS-IEinet	1720	2220
2120	2129	BBCWS-NX	1	Sports Roundup	SP	BBCWS-NXinet	1720	2220
2120	2129	BBCWS-NX	2-6	World Business Report	BE	BBCWS-NXinet	1720	2220
2120	2129	BBCWS-PR	2-6	World Business Report	BE	Sirius141, NPRfm/am, HD+	1720	2220
2120	2128	R.PRG	2	Sports News	SP	R.PRGinet, 5930	1720	2320
2130	2159	ABC-RA	1-5	Breakfast Club	VA	RAinet	1730	0730
2130	2259	ABC-RA	6	Saturday Extra	CA	RAinet	1730	0730
2130	2229	ABC-RN	1-5	Radio National Breakfast	NZ	ABC-RNinet	1730	0730
2130	2259	ABC-RN	6	Saturday Extra	CA	ABC-RNinet	1730	0730
2130	2139	BBCWS-AM	1	The Instant Guide	NA	XM131	1700	2200
2130	2139	BBCWS-AM	2-6	World Briefing	NX	XM131	1730	2230
2130	2159	BBCWS-AM	7	The Interview	PI	XM131	1730	2230
2130	2159	BBCWS-IE	1	Charlie Gillett's World of Music	MZ	BBCWS-IEinet	1700	2200
2130	2149	BBCWS-IE	2-6	Business Daily	BE	BBCWS-IEinet	1730	2230
2130	2159	BBCWS-IE	7	The Interview	PI	BBCWS-IEinet	1730	2230

UTC Time Start	End	Station/ Network	Day(s)	Program Name	Type	Frequncy/ Platform	EDT	Station Time
2130	2159	BBCWS-NX	1	Politics UK	GL	BBCWS-NXinet	1730	2230
2130	2139	BBCWS-NX	2-6	World Briefing	NX	BBCWS-NXinet	1730	2230
2130	2159	BBCWS-NX	7	The Interview	PI	BBCWS-NXinet	1730	2230
2130	2159	BBCWS-PR	2-6	Outlook	GZ	Sirius141, NPRfm/am, HD+	1730	2230
2130	2159	BBCWS-PR	7	The Interview	PI	Sirius141, NPRfm/am, HD+	1730	2230
2130	2229	CBC-R1A	2	As It Happens	NZ	CBAMinet, CBCTinet, CBDinet, CBHAinet, CBIinet, CBNinet, 640, 6160, CBTinet, CBYinet, CBZFinet, CFGBinet	1730	1830
2130	2259	CBC-R1A	3-6	As It Happens	NZ	CBAMinet, CBCTinet, CBDinet, CBHAinet, CBIinet, CBNinet, 640, 6160, CBTinet, CBYinet, CBZFinet, CFGBinet	1730	1830
2130	2159	CBC-R1C	7	White Coat, Black Art	HM	CBWinet, 990	1730	1630
2130	2229	CBC-R1S	2	As It Happens	NA	Sirius137	1730	1730
2130	2259	CBC-R1S	3-6	As It Happens	NA	Sirius137	1730	1730
2130	2154	CRI-RTC	1	China Horizons	DL	CRI-RTCinet	1730	0530
2130	2154	CRI-RTC	2	Frontline	GL	CRI-RTCinet	1730	0530
2130	2154	CRI-RTC	3	Biz China	BE	CRI-RTCinet	1730	0530
2130	2154	CRI-RTC	4	In the Spotlight	DL	CRI-RTCinet	1730	0530
2130	2154	CRI-RTC	5	Voices from Other Lands	DL	CRI-RTCinet	1730	0530
2130	2154	CRI-RTC	6	Life in China	DL	CRI-RTCinet	1730	0530
2130	2154	CRI-RTC	7	Listeners' Garden	LI	CRI-RTCinet	1730	0530
2130	2144	DW	1/7	Sports Report	SP	DWinet	1730	2330
2130	2159	DW	2	Eurovox	AC	DWinet	1730	2330
2130	2159	DW	3	Hits in Germany	MP	DWinet	1730	2330
2130	2159	DW	4	Arts on the Air	AC	DWinet	1730	2330
2130	2159	DW	5	Cool	CS	DWinet	1730	2330
2130	2159	DW	6	Dialogue	CS	DWinet	1730	2330
2130	2229	IRIB	1-7	Koran Reading, News and features	VA	IRIBinet	1730	0100
2130	2159	RNW	1	Reloaded	VA	RNW2inet	1730	2330
2130	2159	RNW	2	Curious Orange	DL	RNW2inet	1730	2330
2130	2159	RNW	3	The State We're In Midweek Edition	CS	RNW2inet	1730	2330
2130	2159	RNW	4	Radio Books	LD	RNW2inet	1730	2330
2130	2159	RNW	5	Earthbeat	EV	RNW2inet	1730	2330
2130	2159	RNW	6	Bridges with Africa	CS	RNW2inet	1730	2330
2130	2159	RNW	7	Curious Orange	DL	RNW2inet	1730	2330
2130	2159	RTE-R1	6	Cuisle na hEalaiona (in Irish Gaelic)	AC	RTE-R1inet	1730	2230
2130	2159	WRN-NA	1-7	Radio Romania International features	VA	Sirius140, WRN-NAinet	1730	2230
2135	2139	RNZI	1-5	News about New Zealand	NX	RNZIinet, 17675	1735	0935
2140	2159	BBCWS-AM	1	Over to You	LI	XM131	1740	2240
2140	2149	BBCWS-AM	2-6	Analysis	NA	XM131	1740	2240
2140	2149	BBCWS-NX	2-6	Analysis	NA	BBCWS-NXinet	1740	2240
2140	2159	RNZI	1-5	RNZI Feature	GI	RNZIinet, 17675	1740	0940
2145	2159	BBC-R4	2-6	A Book at Bedtime	LD	BBC-R4inet	1745	2245
2145	2159	DW	1	Inspired Minds	AC	DWinet	1745	2345
2145	2159	DW	7	Radio D	LL	DWinet	1745	2345
2150	2159	BBCWS-AM	2-6	Sports Roundup	SP	XM131	1750	2250
2150	2159	BBCWS-NX	2-6	Sports Roundup	SP	BBCWS-NXinet	1750	2250
2155	2159	CRI-RTC	1-7	Chinese Studio	LL	CRI-RTCinet	1730	0530
2200	2239	ABC-RA	1-5	AM	NZ	RAinet	1800	0800
2200	2229	ABC-RA	7	Correspondents Report	NA	RAinet	1800	0800
2200	2229	ABC-RN	7	Correspondents Report	NA	ABC-RNinet	1800	0800
2200	2229	BBC-R4	1	The Learning Curve	CS	BBC-R4inet	1800	2300
2200	2229	BBC-R4	2	General feature or documentary series	GI	BBC-R4inet	1800	2300
2200	2229	BBC-R4	3-5	Comedy series	LE	BBC-R4inet	1800	2300
2200	2229	BBC-R4	6	A Good Read	AC	BBC-R4inet	1800	2300
2200	2229	BBC-R4	7	Quiz or panel game	LE	BBC-R4inet	1800	2300
2200	2219	BBCWS-AM	1	World Briefing	NX	XM131	1800	2300
2200	2229	BBCWS-AM	2-6	Outlook	GZ	XM131	1800	2300
2200	2229	BBCWS-AM	7	From Our Own Correspondent	NA	XM131	1800	2300
2200	2259	BBCWS-IE	1	The Strand	AC	BBCWS-IEinet	1800	2300
2200	2229	BBCWS-IE	2-6	Outlook	GZ	BBCWS-IEinet	1800	2300
2200	2229	BBCWS-IE	7	From Our Own Correspondent	NA	BBCWS-IEinet	1800	2300
2200	2219	BBCWS-NX	1-5	World Briefing	NX	BBCWS-NXinet	1800	2300
2200	2229	BBCWS-NX	6	Global Business	BE	BBCWS-NXinet	1800	2300
2200	2229	BBCWS-NX	7	From Our Own Correspondent	NA	BBCWS-NXinet	1800	2300
2200	2219	BBCWS-PR	1-5	World Briefing	NX	Sirius141, NPRfm/am, HD+	1800	2300
2200	2229	BBCWS-PR	6	Global Business	BE	Sirius141, NPRfm/am, HD+	1800	2300
2200	2229	BBCWS-PR	7	From Our Own Correspondent	NA	Sirius141, NPRfm/am, HD+	1800	2300
2200	2229	CBC-R1A	1/7	The World This Weekend	NZ	CBAMinet, CBCTinet, CBDinet, CBHAinet, CBIinet, CBNinet, 640, 6160, CBTinet, CBYinet, CBZFinet, CFGBinet	1800	1900
2200	2259	CBC-R1C	1	Writers and Company	AC	CBWinet, 990	1800	1700
2200	2259	CBC-R1C	7	Regional performance	GI	CBWinet, 990	1800	1700
2200	2229	CBC-R1E	1/7	The World This Weekend	NZ	CBCLinet, CBCSinet, CBEinet, 1550, CBLAinet, CBMEinet, CBOinet, CBQTinet, CBVEinet, CFFBinet	1800	1800
2200	2229	CBC-R1E	2-6	The World At Six	NX	CBCLinet, CBCSinet, CBEinet, 1550, CBLAinet, CBMEinet, CBOinet, CBQTinet, CBVEinet, CFFBinet	1800	1800
2200	2259	CBC-R1M	1	Tapestry	CS	CBKinet, 540, CBRinet, 1010, CBXinet, 740, CFYKinet, CHAKinet	1800	1600

UTC Time Start	End	Station/ Network	Day(s)	Program Name	Type	Frequency/ Platform	EDT	Station Time
2200	2359	CBC-R1M	2-6	Local afternoon program	NZ	CBKinet, 540, CBRinet, 1010, CBXinet, 740, CFYKinet, CHAKinet	1800	1600
2200	2229	CBC-R1M	7	Spark	DX	CBKinet, 540, CBRinet, 1010, CBXinet, 740, CFYKinet, CHAKinet	1800	1600
2200	2259	CBC-R1P	1	Tapestry	CS	CBCVinet, CBTKinet, CBUinet, 690, 6160, CBYGinet, CFWHinet	1800	1500
2200	2229	CBC-R1P	2	Laugh Out Loud	LE	CBCVinet, CBTKinet, CBUinet, 690, 6160, CBYGinet, CFWHinet	1800	1500
2200	2229	CBC-R1P	3	The Choice	GI	CBCVinet, CBTKinet, CBUinet, 690, 6160, CBYGinet, CFWHinet	1800	1500
2200	2229	CBC-R1P	4	The Inside Track	SP	CBCVinet, CBTKinet, CBUinet, 690, 6160, CBYGinet, CFWHinet	1800	1500
2200	2229	CBC-R1P	5	The Age of Persuasion	DX	CBCVinet, CBTKinet, CBUinet, 690, 6160, CBYGinet, CFWHinet	1800	1500
2200	2229	CBC-R1P	6	Festival of Funny	LE	CBCVinet, CBTKinet, CBUinet, 690, 6160, CBYGinet, CFWHinet	1800	1500
2200	2259	CBC-R1P	7	The Next Chapter	AC	CBCVinet, CBTKinet, CBUinet, 690, 6160, CBYGinet, CFWHinet	1800	1500
2200	2229	CBC-R1S	1/7	The World This Weekend	NZ	Sirius137	1800	1800
2200	2259	CBC-RCI	1	The Maple Leaf Mailbag	LI	9610, 9800	1800	1800
2200	2259	CBC-RCI	2-6	The Link (second hour)	GZ	9610, 9800	1800	1800
2200	2259	CBC-RCI	7	Masala Canada	GZ	9610, 9800	1800	1800
2200	2219	CRI-RTC	1/7	News and Reports	NZ	CRI-RTCinet	1800	0600
2200	2254	CRI-RTC	2-6	China Drive	GZ	CRI-RTCinet	1800	0600
2200	2205	DW	1-7	News	NX	DWinet	1800	0000
2200	2209	KBS-WR	1-7	News	NX	KBS-WR2inet	1800	0700
2200	0059	ORF-FM4	7	Digital Konfusion Mixshow (in German and English)	MR	ORF-FM4inet	1800	0000
2200	2229	RNW	1	Network Europe Extra	AC	RNW2inet, WRNna	1800	0000
2200	2229	RNW	2-6	Newsline	NZ	RNW2inet, WRNna	1800	0000
2200	2259	RNW	7	The State We're In	CS	RNW2inet, WRNna	1800	0000
2200	2209	RNZI	1-5	World and Pacific News	GI	RNZIinet, 17675	1800	1000
2200	2211	RTE-R1	1	News and GAA·Sports Results	SP	RTE-R1inet	1800	2300
2200	2244	RTE-R1	2-5	The Late Debate	GL	RTE-R1inet	1800	2300
2200	0059	RTE-R1	6	Late Date	MP	RTE-R1inet	1900	0000
2200	2254	RTE-R1	7	Country Time	MW	RTE-R1inet	1800	2300
2200	2259	RTHK-3	6	Early Show	MP	RTHK-3inet	1800	0600
2200	2359	RTHK-3	7	Sunday Early Show	MP	RTHK-3inet	1800	0600
2200	0159	SABC-SAFM	1-7	SAfm Twilights	MP	SABC-SAFMinet	1800	0000
2200	2259	WRN-NA	1-7	Radio Netherlands Worldwide features	VA	Sirius140, WRN-NAinet	1800	2300
2200	2259	WRS-SUI	1-7	BBC Newshour	NZ	WRS-SUIinet	1800	0000
2205	2229	DW	1-7	Newslink	NZ	DWinet	1805	0005
2206	2229	RNZ-NAT	7	The Sunday Group	NA	RNZ-NATinet	1806	1006
2210	2214	KBS-WR	2-6	News Commentary	NC	KBS-WR2inet	1810	0710
2210	2214	RNZI	1-5	Sports News	SP	RNZIinet, 17675	1810	1010
2212	2254	RTE-R1	1	Balfe's Sunday Best	MV	RTE-R1inet	1812	2312
2215	2259	KBS-WR	1	Korean Pop Interactive	MP	KBS-WR2inet	1815	0715
2215	2244	KBS-WR	2-6	Seoul Calling	NZ	KBS-WR2inet	1815	0715
2215	2259	KBS-WR	7	Worldwide Friendship	LI	KBS-WR2inet	1815	0715
2215	2234	RNZI	1-5	Dateline Pacific	NA	RNZIinet, 17675	1815	1015
2220	2229	BBCWS-AM	1	World Business Report	BE	XM131	1820	2320
2220	2229	BBCWS-NX	1-5	World Business Report	BE	BBCWS-NXinet	1820	2320
2220	2229	BBCWS-PR	1-5	World Business Report	BE	Sirius141, NPRfm/am, HD+	1820	2320
2220	2226	CRI-RTC	1	Reports from Developing Countries	BE	CRI-RTCinet	1820	0620
2220	2226	CRI-RTC	7	CRI Roundup	NX	CRI-RTCinet	1820	0620
2227	2254	CRI-RTC	1/7	China Beat	MP	CRI-RTCinet	1827	0627
2230	2259	ABC-RA	7	Innovations	ST	RAinet	1830	0830
2230	2259	ABC-RN	1	Health Report	HM	ABC-RNinet	1830	0830
2230	2259	ABC-RN	2	Law Report	GL	ABC-RNinet	1830	0830
2230	2259	ABC-RN	3	Rear Vision (from 2/09)	NA	ABC-RNinet	1830	0830
2230	2259	ABC-RN	3	Religion Report (til 1/09)	CS	ABC-RNinet	1830	0830
2230	2259	ABC-RN	4	Future Report (from 2/09)	ST	ABC-RNinet	1830	0830
2230	2259	ABC-RN	4	Media Report (til 1/09)	DX	ABC-RNinet	1830	0830
2230	2259	ABC-RN	5	MovieTime (from 2/09)	AC	ABC-RNinet	1830	0830
2230	2259	ABC-RN	5	The Sports Factor (til 1/09)	SP	ABC-RNinet	1830	0830
2230	2244	ABC-RN	7	Short Story	LD	ABC-RNinet	1830	0830
2230	2259	BBC-R4	1	Something Understood	CS	BBC-R4inet	1830	2330
2230	2259	BBC-R4	2-6	Today in Parliament	GL	BBC-R4inet	1830	2330
2230	2259	BBC-R4	7	Poetry Please	LD	BBC-R4inet	1830	2330
2230	2239	BBCWS-AM	1	World Briefing	NX	XM131	1830	2330
2230	2259	BBCWS-AM	2	Health Check	HM	XM131	1830	2330
2230	2259	BBCWS-AM	3	Digital Planet	DX	XM131	1830	2330
2230	2259	BBCWS-AM	4	Discovery	ST	XM131	1830	2330
2230	2259	BBCWS-AM	5	One Planet	EV	XM131	1830	2330
2230	2259	BBCWS-AM	6	Science in Action	ST	XM131	1830	2330
2230	2259	BBCWS-AM	7	Charlie Gillett's World of Music	MZ	XM131	1830	2330
2230	2259	BBCWS-IE	2-6	The Strand	AC	BBCWS-IEinet	1830	2330
2230	2359	BBCWS-IE	7	The Instant Guide	NA	BBCWS-IEinet	1830	2330
2230	2239	BBCWS-NX	1-5	World Briefing	NX	BBCWS-NXinet	1830	2330
2230	2259	BBCWS-NX	6/7	Politics UK	GL	BBCWS-NXinet	1830	2330

UTC Time Start	End	Station/ Network	Day(s)	Program Name	Type	Frequncy/ Platform	EDT	Station Time
2230	2239	BBCWS-PR	1-5	World Briefing	NX	Sirius141, NPRfm/am, HD+	1830	2330
2230	2259	BBCWS-PR	6	One Planet	EV	Sirius141, NPRfm/am, HD+	1830	2330
2230	2259	BBCWS-PR	7	Heart and Soul	CS	Sirius141, NPRfm/am, HD+	1830	2330
2230	2329	CBC-R1A	1	Dispatches	NA	CBAMinet, CBCTinet, CBDinet, CBHAinet, CBIinet, CBNinet, 640, 6160, CBTinet, CBYinet, CBZFinet, CFGBinet	1830	1930
2230	2329	CBC-R1A	2	Dispatches	NA	CBAMinet, CBCTinet, CBDinet, CBHAinet, CBIinet, CBNinet, 640, 6160, CBTinet, CBYinet, CBZFinet, CFGBinet	1830	1930
2230	2259	CBC-R1A	7	Laugh Out Loud	LE	CBAMinet, CBCTinet, CBDinet, CBHAinet, CBIinet, CBNinet, 640, 6160, CBTinet, CBYinet, CBZFinet, CFGBinet	1830	1930
2230	2329	CBC-R1E	1	Dispatches	NA	CBCLinet, CBCSinet, CBEinet, 1550, CBLAinet, CBMEinet, CBOinet, CBQTinet, CBVEinet, CFFBinet	1830	1830
2230	2329	CBC-R1E	2	As It Happens	NZ	CBCLinet, CBCSinet, CBEinet, 1550, CBLAinet, CBMEinet, CBOinet, CBQTinet, CBVEinet, CFFBinet	1830	1830
2230	2359	CBC-R1E	3-6	As It Happens	NZ	CBCLinet, CBCSinet, CBEinet, 1550, CBLAinet, CBMEinet, CBOinet, CBQTinet, CBVEinet, CFFBinet	1830	1830
2230	2259	CBC-R1E	7	Laugh Out Loud	LE	CBCLinet, CBCSinet, CBEinet, 1550, CBLAinet, CBMEinet, CBOinet, CBQTinet, CBVEinet, CFFBinet	1830	1830
2230	2259	CBC-R1M	7	White Coat, Black Art	HM	CBKinet, 540, CBRinet, 1010, CBXinet, 740, CFYKinet, CHAKinet	1830	1630
2230	2259	CBC-R1S	1	C'est la vie	DL	Sirius137	1830	1830
2230	2259	CBC-R1S	2	The Choice	GI	Sirius137	1830	1830
2230	2259	CBC-R1S	7	The Inside Track	SP	Sirius137	1830	1830
2230	2259	CRI-WASH	1-7	Chinese Writings	LD	CRI-WASHinet	1830	0630
2230	2259	DW	1	A World of Music	MV	DWinet	1830	0030
2230	2259	DW	2	World in Progress	BE	DWinet	1830	0030
2230	2259	DW	3	Spectrum	ST	DWinet	1830	0030
2230	2259	DW	5	Living Planet	EV	DWinet	1830	0030
2230	2259	DW	6	Inside Europe	NZ	DWinet	1830	0230
2230	2259	DW	7	Insight	NA	DWinet	1830	0030
2230	2234	R.PRG	1/7	News and Current Affairs	NX	R.PRGinet, 5930	1830	0030
2230	2239	R.PRG	2-6	News and Current Affairs	NX	R.PRGinet, 5930	1830	0030
2230	2259	RNW	1	Reloaded	VA	RNW2inet, WRNna	1830	0030
2230	2259	RNW	2	Curious Orange	DL	RNW2inet, WRNna	1830	0030
2230	2259	RNW	3	The State We're In Midweek Edition	CS	RNW2inet, WRNna	1830	0030
2230	2259	RNW	4	Radio Books	LD	RNW2inet, WRNna	1830	0030
2230	2259	RNW	5	Earthbeat	EV	RNW2inet, WRNna	1830	0030
2230	2259	RNW	6	Bridges with Africa	CS	RNW2inet, WRNna	1830	0030
2230	2259	RNZ-NAT	7	Hidden Treasures with Trevor Reekie	MZ	RNZ-NATinet	1830	1030
2230	0029	RTHK-3	1-5	Hong Kong Today	NZ	RTHK-3inet	1830	0630
2235	2244	R.PRG	1	Mailbox	LI	R.PRGinet, 5930	1835	0035
2235	2244	R.PRG	7	Magazine	DL	R.PRGinet, 5930	1835	0035
2235	2359	RNZI	1-5	(see RNZ-NAT)		RNZIinet, 15720	1835	1035
2240	2259	ABC-RA	1-5	Breakfast Club	VA	RAinet	1840	0840
2240	2249	BBCWS-AM	1	Analysis	NA	XM131	1840	2340
2240	2259	BBCWS-IE	7	Over to You	LI	BBCWS-IEinet	1820	2320
2240	2249	BBCWS-NX	1-5	Analysis	NA	BBCWS-NXinet	1840	2340
2240	2249	BBCWS-PR	1-5	Analysis	NA	Sirius141, NPRfm/am, HD+	1840	2340
2240	2249	R.PRG	2	One on One	PI	R.PRGinet, 5930	1840	0040
2240	2258	R.PRG	3	Talking Point	DL	R.PRGinet, 5930	1840	0040
2240	2258	R.PRG	4	Czechs in History (monthly)	AC	R.PRGinet, 5930	1840	0040
2240	2258	R.PRG	4	Czechs Today (monthly)	CS	R.PRGinet, 5930	1840	0040
2240	2258	R.PRG	4	Spotlight (fortnightly)	TR	R.PRGinet, 5930	1840	0040
2240	2248	R.PRG	5	Panorama	CS	R.PRGinet, 5930	1840	0040
2240	2248	R.PRG	6	Business News	BE	R.PRGinet, 5930	1840	0040
2245	2259	ABC-RN	7	Ockham's Razor	ST	ABC-RNinet	1845	0845
2245	2259	KBS-WR	2	Faces of Korea	CS	KBS-WR2inet	1845	0745
2245	2259	KBS-WR	3	Business Watch	BE	KBS-WR2inet	1845	0745
2245	2259	KBS-WR	4	Culture on the Move	AC	KBS-WR2inet	1845	0745
2245	2259	KBS-WR	5	Korea Today and Tomorrow	DL	KBS-WR2inet	1845	0745
2245	2259	KBS-WR	6	Seoul Report	DL	KBS-WR2inet	1845	0745
2245	2259	KBS-WR	6	Seoul Report	DL	KBS-WR2inet	1845	0745
2245	2249	R.PRG	1	Letter from Prague	DL	R.PRGinet, 5930	1845	0045
2245	2248	R.PRG	7	Sound Czech	LL	R.PRGinet, 5930	1849	0049
2245	2259	RNZ-NAT	1-5	The Book Reading	LD	RNZ-NATinet	1845	1045
2245	2259	RTE-R1	2-5	Book on One	LD	RTE-R1inet	1845	2345
2249	2258	R.PRG	1	Czech Books (fortnightly)	AC	R.PRGinet, 5930	1849	0049
2249	2258	R.PRG	1	Magic Carpet (monthly)	MZ	R.PRGinet, 5930	1849	0049
2249	2258	R.PRG	1	Music Profile (monthly)	MX	R.PRGinet, 5930	1849	0049
2249	2258	R.PRG	6	The Arts	AC	R.PRGinet, 5930	1849	0049
2249	2258	R.PRG	7	One on One	PI	R.PRGinet, 5930	1849	0049
2250	2259	BBCWS-AM	1	Sports Roundup	BE	XM131	1850	2350
2250	2259	BBCWS-NX	1-5	Sports Roundup	SP	BBCWS-NXinet	1850	2350
2250	2259	BBCWS-PR	1-5	Business Brief	BE	Sirius141, NPRfm/am, HD+	1850	2350
2250	2258	R.PRG	2	Sports News	SP	R.PRGinet, 5930	1850	0050
2255	2259	CRI-RTC	1-7	Music	MX	CRI-RTCinet	1855	0655
2255	2259	RTE-R1	7	Weather and Sea Area Forecast	WX	RTE-R1inet	1855	2355
2300	2359	ABC-RA	1-5	Connect Asia	NZ	RAinet	1900	0900
2300	2329	ABC-RA	6	Asia Review	CA	RAinet	1900	0900

UTC Time Start	End	Station/ Network	Day(s)	Program Name	Type	Frequency/ Platform	EDT	Station Time
2300	2359	ABC-RA	7	Background Briefing	ND	RAinet	1900	0900
2300	2359	ABC-RN	1-5	Life Matters	DL	ABC-RNinet	1900	0900
2300	2359	ABC-RN	6	By Design	AC	ABC-RNinet	1900	0900
2300	2359	ABC-RN	7	Background Briefing	ND	ABC-RNinet	1900	0900
2300	2314	BBC-R4	1	News and Weather	NX	BBC-R4inet	1900	0000
2300	2329	BBC-R4	2-7	News and Weather	NX	BBC-R4inet	1900	0000
2300	2329	BBCWS-AM	1	Documentary feature or series	ND	XM131	1900	0000
2300	2359	BBCWS-AM	2-6	The World Today	NZ	XM131	1900	0000
2300	2329	BBCWS-AM	7	Assignment	ND	XM131	1900	0000
2300	2359	BBCWS-IE	1	The Forum	CA	BBCWS-IEinet	1900	0000
2300	2359	BBCWS-IE	2/4/6	Documentary feature or series	ND	BBCWS-IEinet	1900	0000
2300	2329	BBCWS-IE	3	Global Business	BE	BBCWS-IEinet	1900	0000
2300	2329	BBCWS-IE	5	Assignment	ND	BBCWS-IEinet	1900	0000
2300	2359	BBCWS-IE	7	The Strand	AC	BBCWS-IEinet	1900	0000
2300	2359	BBCWS-NX	1-5	The World Today	NZ	BBCWS-NXinet	1900	0000
2300	2329	BBCWS-NX	6/7	The World Today	NZ	BBCWS-NXinet	1900	0000
2300	2359	BBCWS-PR	1-5	The World Today	NZ	Sirius141, NPRfm/am, HD+	1900	0000
2300	2329	BBCWS-PR	6/7	The World Today	NZ	Sirius141, NPRfm/am, HD+	1900	0000
2300	2344	CBC-R1A	3-6	The Night Time Review	CA	CBAMinet, CBCTinet, CBDinet, CBHAinet, CBIinet, CBNinet, 640, 6160, CBTinet, CBYinet, CBZFinet, CFGBinet	1900	2000
2300	0059	CBC-R1A	7	Vinyl Tap	MP	CBAMinet, CBCTinet, CBDinet, CBHAinet, CBIinet, CBNinet, 640, 6160, CBTinet, CBYinet, CBZFinet, CFGBinet	1900	2000
2300	2329	CBC-R1C	1/7	The World This Weekend	NZ	CBWinet, 990	1900	1800
2300	2329	CBC-R1C	2-6	The World At Six	NX	CBWinet, 990	1900	1800
2300	0059	CBC-R1E	7	Vinyl Tap	MP	CBCLinet, CBCSinet, CBEinet, 1550, CBLAinet, CBMEinet, CBOinet, CBQTinet, CBVEinet, CFFBinet	1900	1900
2300	2359	CBC-R1M	1	Writers and Company	AC	CBKinet, 540, CBRinet, 1010, CBXinet, 740, CFYKinet, CHAKinet	1900	1700
2300	2359	CBC-R1M	7	Regional performance	GI	CBKinet, 540, CBRinet, 1010, CBXinet, 740, CFYKinet, CHAKinet	1900	1700
2300	2329	CBC-R1P	1	WireTap	LE	CBCVinet, CBTKinet, CBUinet, 690, 6160, CBYGinet, CFWHinet	1900	1600
2300	0059	CBC-R1P	2-6	Local afternoon program	NZ	CBCVinet, CBTKinet, CBUinet, 690, 6160, CBYGinet, CFWHinet	1900	1600
2300	2329	CBC-R1P	7	Spark	DX	CBCVinet, CBTKinet, CBUinet, 690, 6160, CBYGinet, CFWHinet	1900	1600
2300	2359	CBC-R1S	1	Dispatches	NA	Sirius137	1900	1900
2300	2359	CBC-R1S	2-6	Ideas	ED	Sirius137	1900	1900
2300	0029	CBC-R1S	7	Go!	LE	Sirius137	1900	1900
2300	2359	CBC-RCI	1	The Maple Leaf Mailbag	LI	Sirius95	1900	1900
2300	0059	CBC-RCI	2-6	The Link	GZ	Sirius95	1900	1900
2300	2359	CBC-RCI	7	Masala Canada	GZ	Sirius95	1900	1900
2300	2359	CRI-ENG	1-5	China Drive	GZ	6040, 11970	1900	0700
2300	2319	CRI-ENG	6/7	News and Reports	NZ	6040, 11970	1900	0700
2300	2329	CRI-RTC	1-7	News and Reports	NZ	CRI-RTCinet	1900	0700
2300	0159	CRI-WASH	1-7	Beyond Beijing	GZ	CRI-WASHinet	1900	0700
2300	2305	DW	1-7	News	NX	DWinet	1900	0100
2300	2309	KBS-WR	1-7	News	NX	KBS-WR1inet	1900	0800
2300	0359	ORF-FM4	2-5	Sleepless (in German and English)	MR	ORF-FM4inet	1900	0100
2300	2329	RNW	1/6	Radio Books	LD	RNW2inet	1900	0100
2300	2329	RNW	2/7	Earthbeat	EV	RNW2inet	1900	0100
2300	2329	RNW	3	Bridges with Africa	CS	RNW2inet	1900	0100
2300	2329	RNW	4	Curious Orange	DL	RNW2inet	1900	0100
2300	2329	RNW	5	Network Europe Extra	AC	RNW2inet	1900	0100
2300	0059	RTE-R1	7-5	Late Date	MP	RTE-R1inet	1900	0000
2300	2329	RTHK-3	6	Today at Seven	NZ	RTHK-3inet	1900	0700
2300	2329	WRN-NA	1-7	Voice of Russia World Service features	VA	Sirius140, WRN-NAinet	1900	0000
2300	0359	WRS-SUI	1-7	Relay BBC World Service	VA	WRS-SUIinet	1900	0100
2305	0004	CBC-RCI	1	The Maple Leaf Mailbag	LI	6100, 9755, RCIinet	1905	1905
2305	0004	CBC-RCI	2-6	The Link (first hour)	GZ	6100, 9755, RCIinet	1905	1905
2305	0004	CBC-RCI	7	Masala Canada	GZ	6100, 9755, RCIinet	1905	1905
2305	2329	DW	1-7	Newslink	NZ	DWinet	1905	0105
2305	2309	RAI-INT	7-5	News Bulletin	NX	RAI-INTinet	1905	0105
2305	2359	RNZ-NAT	7	Ideas	ED	RNZ-NATinet	1905	1105
2310	2314	KBS-WR	2-6	News Commentary	NC	KBS-WR1inet	1910	0810
2315	2344	BBC-R4	1	Thinking Allowed	ED	BBC-R4inet	1915	0015
2315	2359	KBS-WR	1	Korean Pop Interactive	MP	KBS-WR1inet	1915	0815
2315	2344	KBS-WR	2-6	Seoul Calling	NZ	KBS-WR1inet	1915	0815
2315	2359	KBS-WR	7	Worldwide Friendship	LI	KBS-WR1inet	1915	0815
2320	2326	CRI-ENG	6	CRI Roundup	NX	6040, 11970	1920	0720
2320	2326	CRI-ENG	7	Report from Developing Countries	BE	6040, 11970	1920	0720
2327	2351	CRI-ENG	6	Music Memories	MF	6040, 11970	1927	0727
2327	2351	CRI-ENG	7	China Beat	MP	6040, 11970	1927	0727
2330	2359	ABC-RA	6	Talking Point	CA	RAinet	1930	0930
2330	2347	BBC-R4	2-6	Book of the Week	LD	BBC-R4inet	1930	0030
2330	2347	BBC-R4	7	The Late Story	LD	BBC-R4inet	1930	0030
2330	2359	BBCWS-AM	1	Charlie Gillett's World of Music	MZ	XM131	1930	0030
2330	2359	BBCWS-AM	7	Reporting Religion	CS	XM131	1930	0030

UTC Time Start	End	Station/ Network	Day(s)	Program Name	Type	Frequncy/ Platform	EDT	Station Time
2330	2359	BBCWS-IE	2	Health Check	HM	BBCWS-IEinet	1930	0030
2330	2359	BBCWS-IE	3	Digital Planet	DX	BBCWS-IEinet	1930	0030
2330	2359	BBCWS-IE	4	Discovery	ST	BBCWS-IEinet	1930	0030
2330	2359	BBCWS-IE	5	One Planet	EV	BBCWS-IEinet	1930	0030
2330	2359	BBCWS-IE	6	Science in Action	ST	BBCWS-IEinet	1930	0030
2330	2359	BBCWS-NX	6	The Interview	PI	BBCWS-NXinet	1930	0030
2330	2359	BBCWS-NX	7	Reporting Religion	CS	BBCWS-NXinet	1930	0030
2330	2359	BBCWS-PR	6	The Interview	PI	Sirius141, NPRfm/am, HD+	1930	0030
2330	2359	BBCWS-PR	7	Reporting Religion	CS	Sirius141, NPRfm/am, HD+	1930	0030
2330	2359	CBC-R1A	1	C'est la vie	DL	CBAMinet, CBCTinet, CBDinet, CBHAinet, CBIinet, CBNinet, 640, 6160, CBTinet, CBYinet, CBZFinet, CFGBinet	1930	2030
2330	2359	CBC-R1A	1	The Debaters	LE	CBAMinet, CBCTinet, CBDinet, CBHAinet, CBIinet, CBNinet, 640, 6160, CBTinet, CBYinet, CBZFinet, CFGBinet	1930	2030
2330	0029	CBC-R1C	1	Dispatches	NA	CBWinet, 990	1930	1830
2330	0029	CBC-R1C	2	As It Happens	NZ	CBWinet, 990	1930	1830
2330	0059	CBC-R1C	3-6	As It Happens	NZ	CBWinet, 990	1930	1830
2330	2359	CBC-R1C	7	Laugh Out Loud	LE	CBWinet, 990	1930	1830
2330	2359	CBC-R1E	1	C'est la vie	DL	CBCLinet, CBCSinet, CBEinet, 1550, CBLAinet, CBMEinet, CBOinet, CBQTinet, CBVEinet, CFFBinet	1930	1930
2330	0029	CBC-R1E	2	Dispatches	NA	CBCLinet, CBCSinet, CBEinet, 1550, CBLAinet, CBMEinet, CBOinet, CBQTinet, CBVEinet, CFFBinet	1930	1930
2330	2359	CBC-R1P	1	The Inside Track	SP	CBCVinet, CBTKinet, CBUinet, 690, 6160, CBYGinet, CFWHinet	1930	1630
2330	2359	CBC-R1P	7	White Coat, Black Art	HM	CBCVinet, CBTKinet, CBUinet, 690, 6160, CBYGinet, CFWHinet	1930	1630
2330	2354	CRI-RTC	1-5	People in the Know	GL	CRI-RTCinet	1930	0730
2330	2354	CRI-RTC	6	Listeners' Garden	LI	CRI-RTCinet	1930	0730
2330	2354	CRI-RTC	7	China Horizons	DL	CRI-RTCinet	1930	0730
2330	2344	DW	1/7	Sports Report	SP	DWinet	1930	0130
2330	2359	DW	2	Eurovox	AC	DWinet	1930	0130
2330	2359	DW	3	Hits in Germany	MP	DWinet	1930	0130
2330	2259	DW	4	Money Talks	BE	DWinet	1830	0130
2330	2359	DW	4	Arts on the Air	AC	DWinet	1930	0130
2330	2359	DW	5	Cool	CS	DWinet	1930	0130
2330	2359	DW	6	Dialogue	CS	DWinet	1930	0130
2330	2334	R.PRG	1/7	News and Current Affairs	NX	R.PRGinet, 5930, 7345	1930	0130
2330	2339	R.PRG	2-6	News and Current Affairs	NX	R.PRGinet, 5930, 7345	1930	0130
2330	2359	RNW	1	Reloaded	VA	RNW2inet	1930	0130
2330	2359	RNW	2/7	Curious Orange	DL	RNW2inet	1930	0130
2330	2359	RNW	3	The State We're In Midweek Edition	CS	RNW2inet	1930	0130
2330	2359	RNW	4	Radio Books	LD	RNW2inet	1930	0130
2330	2359	RNW	5	Earthbeat	EV	RNW2inet	1930	0130
2330	2359	RNW	6	Bridges with Africa	CS	RNW2inet	1930	0130
2330	2344	RTHK-3	6	Reflections from Asia	NA	RTHK-3inet	1930	0730
2330	2359	WRN-NA	1-7	Israel Radio: News	NX	Sirius140, WRN-NAinet	1930	0030
2335	2344	R.PRG	1	Mailbox	LI	R.PRGinet, 5930, 7345	1935	0135
2335	2344	R.PRG	7	Magazine	DL	R.PRGinet, 5930, 7345	1935	0135
2340	2349	R.PRG	2	One on One	PI	R.PRGinet, 5930, 7345	1940	0140
2340	2358	R.PRG	3	Talking Point	DL	R.PRGinet, 5930, 7345	1940	0140
2340	2358	R.PRG	4	Czechs in History (monthly)	AC	R.PRGinet, 5930, 7345	1940	0140
2340	2358	R.PRG	4	Czechs Today (monthly)	CS	R.PRGinet, 5930, 7345	1940	0140
2340	2358	R.PRG	4	Spotlight (fortnightly)	TR	R.PRGinet, 5930, 7345	1940	0140
2340	2348	R.PRG	5	Panorama	CS	R.PRGinet, 5930, 7345	1940	0140
2340	2348	R.PRG	6	Business News	BE	R.PRGinet, 5930, 7345	1940	0140
2345	2347	BBC-R4	1	Bells on Sunday	MX	BBC-R4inet	1945	0045
2345	2359	CBC-R1A	3-6	Outfront	DL	CBAMinet, CBCTinet, CBDinet, CBHAinet, CBIinet, CBNinet, 640, 6160, CBTinet, CBYinet, CBZFinet, CFGBinet	1945	2045
2345	2359	DW	1	Inspired Minds	AC	DWinet	1945	0145
2345	2359	DW	7	Radio D	LL	DWinet	1945	0145
2345	2359	KBS-WR	2	Faces of Korea	CS	KBS-WR1inet	1945	0845
2345	2359	KBS-WR	3	Business Watch	BE	KBS-WR1inet	1945	0845
2345	2359	KBS-WR	4	Culture on the Move	AC	KBS-WR1inet	1945	0845
2345	2359	KBS-WR	5	Korea Today and Tomorrow	DL	KBS-WR1inet	1945	0845
2345	2359	KBS-WR	6	Seoul Report	DL	KBS-WR1inet	1945	0845
2345	2349	R.PRG	1	Letter from Prague	DL	R.PRGinet, 5930, 7345	1945	0145
2345	2348	R.PRG	7	Sound Czech	LL	R.PRGinet, 5930, 7345	1949	0149
2345	2359	RTHK-3	6	Hong Kong Heritage	AC	RTHK-3inet	1945	0745
2348	2359	BBC-R4	1-7	Shipping Forecast	WX	BBC-R4inet	1948	0048
2349	2358	R.PRG	1	Czech Books (fortnightly)	AC	R.PRGinet, 5930, 7345	1949	0149
2349	2358	R.PRG	1	Magic Carpet (monthly)	MZ	R.PRGinet, 5930, 7345	1949	0149
2349	2358	R.PRG	1	Music Profile (monthly)	MX	R.PRGinet, 5930, 7345	1949	0149
2349	2358	R.PRG	6	The Arts	AC	R.PRGinet, 5930, 7345	1949	0149
2349	2358	R.PRG	7	One on One	PI	R.PRGinet, 5930, 7345	1949	0149
2350	2358	R.PRG	2	Sports News	SP	R.PRGinet, 5930, 7345	1950	0150
2350	2358	R.PRG	2	Sports News	SP	R.PRGinet, 5930, 7345	1950	0150
2352	2357	CRI-ENG	6/7	Learning Chinese Now	LL	6040, 11970	1952	0752
2355	2359	CRI-RTC	1-7=	Music	DL	CRI-RTCinet	1930	0730

Classified Program Lists

The following lists classify programs into popular topic areas to provide you with a handy cross-reference to the information in the Worldwide Listening Guide Program List, as well as an easy way to find programs of interest to you. Within each classification, the programs are arranged first by broadcast station and then in alphabetical order by program title. Each listing indicates station, program title, and the days and times the program is broadcast. All times are in UTC.

Once you identify a program that you wish to listen to, use the UTC times listed to find the program in the main Worldwide Listening Guide Program List to determine the

broadcast frequencies for the selected broadcast. You also can use this listing to find out the times of repeat broadcasts of programs you like, thus identifying a time for listening that is more convenient for you.

Finally, some of these programs may be available as on-demand files on the Internet. Combine this reference with others in this Guide, such as the information on station and audio sites on the Internet, to see if a station has added this on-demand feature for a selected program or even begun to archive past broadcasts of the program to allow you to hear again what you liked, or hear for the first time what you may have missed.

Arts, Culture & History Programs

Time	Station/Network	Program Name	Day(s)
1530	BBC-R4	A Good Read	3
2200	BBC-R4	A Good Read	6
0320	VOR-WS	A Stroll Around the Kremlin	2
0330	VOR-WS	A Stroll Around the Kremlin	1/4/7
0830	VOR-WS	A Stroll Around the Kremlin	1/4
0930	VOR-WS	A Stroll Around the Kremlin	2
1530	RNZ-NAT	An Author's View	3
1900	SABC-SAFM	Art Matters	6
0530	DW	Arts on the Air	4
0730	DW	Arts on the Air	4
0930	DW	Arts on the Air	4
1130	DW	Arts on the Air	4
1330	DW	Arts on the Air	4
1530	DW	Arts on the Air	4
1730	DW	Arts on the Air	4
1930	DW	Arts on the Air	4
2130	DW	Arts on the Air	4
2330	DW	Arts on the Air	4
1030	BBC-R4	Arts or Drama feature or series	3
0200	ABC-RA	Artworks	1
0500	ABC-RA	Artworks	1
0734	ABC-RA	Artworks	7
1430	ABC-RA	Artworks	6
0000	ABC-RN	Artworks	1
0500	ABC-RN	Artworks	3
1500	ABC-RN	Artworks	1
1035	ABC-RN	Artworks Feature	2
1000	ABC-RN	Artworks Feature (from 2/09)	1
0106	RNZ-NAT	At the Movies	1
0730	RNZ-NAT	At the Movies	4
1230	RNZ-NAT	At the Movies	2
1700	WRS-SUI	Bookmark (2nd Sat. of the month)	7
0930	WRS-SUI	Bookmark (2nd Tues. of the month)	3
0225	RNZ-NAT	Books	1
0500	ABC-RN	By Design	4
2300	ABC-RN	By Design	6
2130	RTE-R1	Cuisle na hEalaiona (in Irish Gaelic)	6
1440	CRI-WASH	Cultural Carousel	3-7
0240	CRI-WASH	Cultural Horizons	4-1
0245	KBS-WR	Culture on the Move	4
0345	KBS-WR	Culture on the Move	4
0515	KBS-WR	Culture on the Move	4
0645	KBS-WR	Culture on the Move	4
0815	KBS-WR	Culture on the Move	4
0915	KBS-WR	Culture on the Move	4

Arts, Culture & History Programs *continued*

Time	Station/Network	Program Name	Day(s)
1145	KBS-WR	Culture on the Move	4
1215	KBS-WR	Culture on the Move	4
1245	KBS-WR	Culture on the Move	4
1515	KBS-WR	Culture on the Move	4
2045	KBS-WR	Culture on the Move	4
2245	KBS-WR	Culture on the Move	4
2345	KBS-WR	Culture on the Move	4
0119	R.PRG	Czech Books (fortnightly)	2
0219	R.PRG	Czech Books (fortnightly)	2
0419	R.PRG	Czech Books (fortnightly)	2
0449	R.PRG	Czech Books (fortnightly)	2
0819	R.PRG	Czech Books (fortnightly)	1
1149	R.PRG	Czech Books (fortnightly)	1
1419	R.PRG	Czech Books (fortnightly)	1
1719	R.PRG	Czech Books (fortnightly)	1
1819	R.PRG	Czech Books (fortnightly)	1
2119	R.PRG	Czech Books (fortnightly)	1
2249	R.PRG	Czech Books (fortnightly)	1
2349	R.PRG	Czech Books (fortnightly)	1
0110	R.PRG	Czechs in History (monthly)	5
0210	R.PRG	Czechs in History (monthly)	5
0410	R.PRG	Czechs in History (monthly)	5
0440	R.PRG	Czechs in History (monthly)	5
0810	R.PRG	Czechs in History (monthly)	4
1140	R.PRG	Czechs in History (monthly)	4
1410	R.PRG	Czechs in History (monthly)	4
1710	R.PRG	Czechs in History (monthly)	4
1810	R.PRG	Czechs in History (monthly)	4
2110	R.PRG	Czechs in History (monthly)	4
2240	R.PRG	Czechs in History (monthly)	4
2340	R.PRG	Czechs in History (monthly)	4
0400	VOR-WS	Encyclopedia "All Russia"	3/6
0500	VOR-WS	Encyclopedia "All Russia"	1/2
0700	VOR-WS	Encyclopedia "All Russia"	4/7
0530	DW	Eurovox	2
0730	DW	Eurovox	2
0930	DW	Eurovox	2
1130	DW	Eurovox	2
1330	DW	Eurovox	2
1530	DW	Eurovox	2
1730	DW	Eurovox	2
1930	DW	Eurovox	2
2130	DW	Eurovox	2
2330	DW	Eurovox	2
2100	RTE-R1	Feature/Arts/Documentary series	2-5

Time	Station/Network	Program Name	Day(s)
0355	PR-EXT	Focus	2/6
0730	PR-EXT	Focus	6
1205	PR-EXT	Focus	2
1205	PR-EXT	Focus	5
1725	PR-EXT	Focus	5
1955	PR-EXT	Focus	5
1815	BBC-R4	Front Row	2-6
0300	CBC-R1A	Global Arts and Entertainment	1
0500	CBC-R1C	Global Arts and Entertainment	2
0400	CBC-R1E	Global Arts and Entertainment	2
0600	CBC-R1M	Global Arts and Entertainment	2
0700	CBC-R1P	Global Arts and Entertainment	2
1400	ABC-RA	Hindsight	5
0300	ABC-RN	Hindsight	5
0400	ABC-RN	Hindsight	1
1600	ABC-RN	Hindsight	5
1230	RNZ-NAT	History Repeated	7
1000	BBC-R4	History series	4
1015	RTHK-3	Hong Kong Heritage	1
2345	RTHK-3	Hong Kong Heritage	6
0145	DW	Inspired Minds	1
1945	DW	Inspired Minds	1
2145	DW	Inspired Minds	1
2345	DW	Inspired Minds	1
0600	ABC-RN	Life and Times	7
2000	ABC-RN	Life and Times	7
0545	ABC-RN	Lingua Franca	5/7
1400	BBC-R4	Making History	3
0230	VOR-WS	Moscow Yesterday and Today	7
0330	VOR-WS	Moscow Yesterday and Today	5
0430	VOR-WS	Moscow Yesterday and Today	2
0530	VOR-WS	Moscow Yesterday and Today	4
0630	VOR-WS	Moscow Yesterday and Today	5
0830	VOR-WS	Moscow Yesterday and Today	6
0330	ABC-RA	MovieTime	1
0630	ABC-RA	MovieTime	1
1400	ABC-RA	MovieTime	6
0130	ABC-RN	MovieTime	1
1800	ABC-RN	MovieTime	5
1030	ABC-RA	MovieTime (from 2/09)	6
2230	ABC-RN	MovieTime (from 2/09)	5
0900	ABC-RN	MovieTime (til 1/09)	5
1530	ABC-RA	MovieTime (til 2/09)	6
0330	PR-EXT	Network Europe Extra	2
0705	PR-EXT	Network Europe Extra	4
1700	PR-EXT	Network Europe Extra	1
1930	PR-EXT	Network Europe Extra	1
0730	RFI	Network Europe Extra	2
1430	RFI	Network Europe Extra	2
0000	RNW	Network Europe Extra	2/6
0100	RNW	Network Europe Extra	6
0200	RNW	Network Europe Extra	2/6
0300	RNW	Network Europe Extra	2
0400	RNW	Network Europe Extra	2
0405	RNW	Network Europe Extra	2
0500	RNW	Network Europe Extra	2
0500	RNW	Network Europe Extra	6
0505	RNW	Network Europe Extra	2
0600	RNW	Network Europe Extra	6
0605	RNW	Network Europe Extra	2
0700	RNW	Network Europe Extra	2/6
0705	RNW	Network Europe Extra	2
0800	RNW	Network Europe Extra	6
0805	RNW	Network Europe Extra	2
0900	RNW	Network Europe Extra	1
0930	RNW	Network Europe Extra	5
1000	RNW	Network Europe Extra	1
1030	RNW	Network Europe Extra	5
1130	RNW	Network Europe Extra	5
1200	RNW	Network Europe Extra	1
1230	RNW	Network Europe Extra	5
1305	RNW	Network Europe Extra	1
1400	RNW	Network Europe Extra	5
1500	RNW	Network Europe Extra	1
1605	RNW	Network Europe Extra	1

Time	Station/Network	Program Name	Day(s)
1630	RNW	Network Europe Extra	5
1705	RNW	Network Europe Extra	1
1830	RNW	Network Europe Extra	5
1905	RNW	Network Europe Extra	1
2030	RNW	Network Europe Extra	5
2100	RNW	Network Europe Extra	1
2200	RNW	Network Europe Extra	1
2300	RNW	Network Europe Extra	5
1530	RNZ-NAT	New Zealand Books	2
0330	RTE-R1	Off the Shelf	2
1800	RTE-R1	Off the Shelf	7
1630	RNZ-NAT	On Screen	7
1500	BBC-R4	Open Book	1/5
0315	SABC-CHAF	Our Heritage	5
0615	SABC-CHAF	Our Heritage	1
0715	SABC-CHAF	Our Heritage	5
1415	SABC-CHAF	Our Heritage	1
0830	BBC-R4	Phill Jupitus' Strips	3
0100	CBC-R1A	Q	3-7
1300	CBC-R1A	Q	2-5
1300	CBC-R1A	Q	6
0300	CBC-R1C	Q	3-7
1500	CBC-R1C	Q	2-5
1500	CBC-R1C	Q	6
0200	CBC-R1E	Q	3-7
1400	CBC-R1E	Q	2-5
1400	CBC-R1E	Q	6
0400	CBC-R1M	Q	3-7
1600	CBC-R1M	Q	2-5
1600	CBC-R1M	Q	6
0500	CBC-R1P	Q	3-7
1700	CBC-R1P	Q	2-5
1700	CBC-R1P	Q	6
0300	CBC-R1S	Q	3-7
1300	CBC-R1S	Q	2-6
1900	CBC-R1S	Q	2-6
1815	BBC-R4	Saturday Review	7
0800	RNZ-NAT	Sounds Historical	1
0800	BBC-R4	Start the Week	2
0340	PR-EXT	Studio 15	5
0715	PR-EXT	Studio 15	5
1230	PR-EXT	Studio 15	4
1600	BBCWS-AM	Sunday Sportsworld	1
0900	WRS-SUI	Swiss By Design	7
0700	WRS-SUI	Swiss by Design	5
0220	RTI	Taiwan Indie	6
0119	R.PRG	The Arts	7
0219	R.PRG	The Arts	7
0419	R.PRG	The Arts	7
0449	R.PRG	The Arts	7
0819	R.PRG	The Arts	6
1149	R.PRG	The Arts	6
1419	R.PRG	The Arts	6
1719	R.PRG	The Arts	6
1819	R.PRG	The Arts	6
2119	R.PRG	The Arts	6
2249	R.PRG	The Arts	6
2349	R.PRG	The Arts	6
0040	RNZ-NAT	The Arts on Sunday with Lynn Freeman	1
1900	RTE-R1	The Arts Show	2-6
0000	ABC-RN	The Book Show	2-6
1000	ABC-RN	The Book Show (from 2/09)	2-5
1400	ABC-RN	The Book Show (from 2/09)	1
0400	RTE-R1	The Dave Fanning Show	7
1800	RTE-R1	The Dave Fanning Show	2-6
1530	BBC-R4	The Film Programme	6
0800	ABC-RA	The Margaret Throsby Interview	7
1600	ABC-RA	The Margaret Throsby Interview	7
1800	CBC-R1A	The Next Chapter	7
2000	CBC-R1C	The Next Chapter	7
1900	CBC-R1E	The Next Chapter	7
2100	CBC-R1M	The Next Chapter	7
2200	CBC-R1P	The Next Chapter	7
1000	CBC-R1S	The Next Chapter	7
1700	CBC-R1S	The Next Chapter	7

Arts, Culture & History Programs *continued*

Time	Station/Network	Program Name	Day(s)
0000	BBCWS-AM	The Strand	1
0500	BBCWS-AM	The Strand	1
1300	BBCWS-AM	The Strand	7
1930	BBCWS-AM	The Strand	2-6
0130	BBCWS-IE	The Strand	3-7
0800	BBCWS-IE	The Strand	7
0930	BBCWS-IE	The Strand	2-6
1400	BBCWS-IE	The Strand	1
1430	BBCWS-IE	The Strand	2-6
1830	BBCWS-IE	The Strand	2-6
1900	BBCWS-IE	The Strand	7
2200	BBCWS-IE	The Strand	1
2230	BBCWS-IE	The Strand	2-6
2300	BBCWS-IE	The Strand	7
0600	BBCWS-PR	The Strand	7
0830	BBCWS-PR	The Strand	2-6
1030	BBCWS-PR	The Strand	2-6
1500	BBCWS-PR	The Strand	7
1600	BBCWS-PR	The Strand	1
0230	VOR-WS	The VOR Treasure Store	5
0530	VOR-WS	The VOR Treasure Store	2/6
0630	VOR-WS	The VOR Treasure Store	1
0830	VOR-WS	The VOR Treasure Store	5
0930	VOR-WS	The VOR Treasure Store	4
1530	RNZ-NAT	The Word	4
0210	RTI	Time Traveler	1
1035	ABC-RA	Verbatim	7
1900	ABC-RN	Verbatim	7
0900	ABC-RN	Verbatim (from 2/09)	6
1035	ABC-RN	Verbatim (til 1/09)	4
0200	CBC-R1A	Writers and Company	6
1800	CBC-R1A	Writers and Company	1
0400	CBC-R1C	Writers and Company	6
2200	CBC-R1C	Writers and Company	1
0300	CBC-R1E	Writers and Company	6
1900	CBC-R1E	Writers and Company	1
0500	CBC-R1M	Writers and Company	6
2300	CBC-R1M	Writers and Company	1
0000	CBC-R1P	Writers and Company	2
0600	CBC-R1P	Writers and Company	6
0000	CBC-R1S	Writers and Company	6
0800	CBC-R1S	Writers and Company	6
1000	CBC-R1S	Writers and Company	5
1500	CBC-R1S	Writers and Company	1

Business, Finance & Economic Development Programs

Time	Station/Network	Program Name	Day(s)
0330	ABC-RA	Asia Pacific Business	7
0630	ABC-RA	Asia Pacific Business	7
1000	ABC-RA	Asia Pacific Business	7
0550	WRS-SUI	BBC World Business Report	2-6
0027	CRI-ENG	Biz China	3
0127	CRI-ENG	Biz China	3
0327	CRI-ENG	Biz China	3
0427	CRI-ENG	Biz China	3
0527	CRI-ENG	Biz China	3
0627	CRI-ENG	Biz China	3
1327	CRI-ENG	Biz China	2
1427	CRI-ENG	Biz China	2
1527	CRI-ENG	Biz China	2
0030	CRI-RTC	Biz China	3
0130	CRI-RTC	Biz China	3
0230	CRI-RTC	Biz China	3
0330	CRI-RTC	Biz China	3
0430	CRI-RTC	Biz China	3
0530	CRI-RTC	Biz China	3
0630	CRI-RTC	Biz China	3
0830	CRI-RTC	Biz China	3
0930	CRI-RTC	Biz China	3
1230	CRI-RTC	Biz China	3

Business, Finance & Economic Development Programs *continued*

Time	Station/Network	Program Name	Day(s)
1330	CRI-RTC	Biz China	3
1430	CRI-RTC	Biz China	3
1530	CRI-RTC	Biz China	3
1630	CRI-RTC	Biz China	3
1830	CRI-RTC	Biz China	3
1930	CRI-RTC	Biz China	3
2030	CRI-RTC	Biz China	3
2130	CRI-RTC	Biz China	3
1300	CRI-WASH	Biz China	3
1330	CRI-WASH	Biz China	4
2000	CRI-WASH	Biz China	4
0050	BBCWS-PR	Business Brief	2-6
2250	BBCWS-PR	Business Brief	1-5
0730	BBCWS-AM	Business Daily	2-6
1140	BBCWS-AM	Business Daily	2-6
0730	BBCWS-IE	Business Daily	2-6
1140	BBCWS-IE	Business Daily	2-6
2130	BBCWS-IE	Business Daily	2-6
0730	BBCWS-NX	Business Daily	2-6
0140	BBCWS-PR	Business Daily	2-6
0440	BBCWS-PR	Business Daily	2-6
1140	BBCWS-PR	Business Daily	2-6
1430	BBCWS-PR	Business Daily	2-6
1940	BBCWS-PR	Business Daily	2-6
0110	R.PRG	Business News	7
0210	R.PRG	Business News	7
0410	R.PRG	Business News	7
0440	R.PRG	Business News	7
0810	R.PRG	Business News	6
1140	R.PRG	Business News	6
1410	R.PRG	Business News	6
1710	R.PRG	Business News	6
1810	R.PRG	Business News	6
2110	R.PRG	Business News	6
2240	R.PRG	Business News	6
2340	R.PRG	Business News	6
0245	KBS-WR	Business Watch	3
0345	KBS-WR	Business Watch	3
0515	KBS-WR	Business Watch	3
0645	KBS-WR	Business Watch	3
0815	KBS-WR	Business Watch	3
0915	KBS-WR	Business Watch	3
1145	KBS-WR	Business Watch	3
1215	KBS-WR	Business Watch	3
1245	KBS-WR	Business Watch	3
1515	KBS-WR	Business Watch	3
2045	KBS-WR	Business Watch	3
2245	KBS-WR	Business Watch	3
2345	KBS-WR	Business Watch	3
0350	PR-EXT	Business Week	7
1205	PR-EXT	Business Week	6
1720	PR-EXT	Business Week	6
1950	PR-EXT	Business Week	6
0930	BBCWS-AM	Business Weekly	7
1930	BBCWS-AM	Business Weekly	7
0030	BBCWS-IE	Business Weekly	7
0430	BBCWS-IE	Business Weekly	1
0230	BBCWS-PR	Business Weekly	1
0930	BBCWS-PR	Business Weekly	1
1630	BBCWS-PR	Business Weekly	7
1830	BBCWS-PR	Business Weekly	1
1200	CRI-WASH	China Biz Report	1-7
0400	RTE-R1	Farm Week	2
2100	RTE-R1	Farm Week	6
0445	BBC-R4	Farming Today	2-6
0535	BBC-R4	Farming Today This Week	7
0315	SABC-CHAF	Gateway to Africa	6
0000	BBCWS-AM	Global Business	4
0300	BBCWS-AM	Global Business	2
0500	BBCWS-AM	Global Business	4
1030	BBCWS-AM	Global Business	7
1400	BBCWS-AM	Global Business	3
1900	BBCWS-AM	Global Business	3

Time	Station/Network	Program Name	Day(s)
0100	BBCWS-IE	Global Business	1
0900	BBCWS-IE	Global Business	3
1200	BBCWS-IE	Global Business	3
1900	BBCWS-IE	Global Business	3
2300	BBCWS-IE	Global Business	3
0130	BBCWS-NX	Global Business	7
0830	BBCWS-NX	Global Business	7
1300	BBCWS-NX	Global Business	1
2200	BBCWS-NX	Global Business	6
0130	BBCWS-PR	Global Business	7
0430	BBCWS-PR	Global Business	7
0730	BBCWS-PR	Global Business	6
0830	BBCWS-PR	Global Business	7
1000	BBCWS-PR	Global Business	7
1300	BBCWS-PR	Global Business	1
1830	BBCWS-PR	Global Business	6
2200	BBCWS-PR	Global Business	6
1630	RNZ-NAT	Global Business	5
1930	BBC-R4	In Business	5
2030	BBC-R4	In Business	1
1600	SABC-SAFM	Market Update	2-6
1100	BBC-R4	Money Box	7
2000	BBC-R4	Money Box	1
1400	BBC-R4	Money Box Live	2
0030	DW	Money Talks	5
0430	DW	Money Talks	4
0630	DW	Money Talks	4
0830	DW	Money Talks	4
1030	DW	Money Talks	4
1230	DW	Money Talks	4
1430	DW	Money Talks	4
1630	DW	Money Talks	4
1830	DW	Money Talks	4
2030	DW	Money Talks	4
2330	DW	Money Talks	4
0515	SABC-CHAF	NEPAD Focus	2
1935	RNZI	Pacific Business Report	1-5
2055	RNZI	Pacific Business Report	1-5
2320	CRI-ENG	Report from Developing Countries	7
0020	CRI-ENG	Reports from Developing Countries	1
0120	CRI-ENG	Reports from Developing Countries	1
0320	CRI-ENG	Reports from Developing Countries	1
0420	CRI-ENG	Reports from Developing Countries	1
0520	CRI-ENG	Reports from Developing Countries	1
0620	CRI-ENG	Reports from Developing Countries	1
1120	CRI-ENG	Reports from Developing Countries	1
1320	CRI-ENG	Reports from Developing Countries	7
1420	CRI-ENG	Reports from Developing Countries	7
1520	CRI-ENG	Reports from Developing Countries	7
0720	CRI-RTC	Reports from Developing Countries	1
1020	CRI-RTC	Reports from Developing Countries	1
1120	CRI-RTC	Reports from Developing Countries	1
1720	CRI-RTC	Reports from Developing Countries	1
2220	CRI-RTC	Reports from Developing Countries	1
0715	SABC-CHAF	SADC Calling	7
1715	SABC-CHAF	SADC Calling	1
0400	RTE-R1	Seascapes	6
2250	BBCWS-AM	Sports Roundup	1
1000	WRS-SUI	Survival Guide	7
0240	RTI	Taiwan Outlook	6
0615	SABC-CHAF	Tam Tam Express	7
0405	PR-EXT	The Biz	4
0730	PR-EXT	The Biz	4
1230	PR-EXT	The Biz	3
1730	PR-EXT	The Biz	3
2000	PR-EXT	The Biz	3
1630	BBC-R4	The Bottom Line	7
0900	RTE-R1	The Business	1
1113	CBC-R1S	The Business Network, Sports & Information	2-6
0930	BBCWS-NX	The Interview	7
0530	BBCWS-AM	The Strand	3-7
0300	RTE-R1	This Business	2
0330	RNZI	Tradewinds	4

Time	Station/Network	Program Name	Day(s)
0730	RNZI	Tradewinds	4
1130	RNZI	Tradewinds	3
1330	RNZI	Tradewinds	3
1630	RNZI	Tradewinds	3
1708	RNZI	Tradewinds	3
2115	RNZI	Tradewinds	3
0120	BBCWS-AM	World Business Report	2-6
0420	BBCWS-AM	World Business Report	2-6
0820	BBCWS-AM	World Business Report	2-6
1120	BBCWS-AM	World Business Report	2-6
1540	BBCWS-AM	World Business Report	2-6
1820	BBCWS-AM	World Business Report	2-6
2120	BBCWS-AM	World Business Report	2-6
2220	BBCWS-AM	World Business Report	1
0120	BBCWS-IE	World Business Report	2-6
0420	BBCWS-IE	World Business Report	2-6
0820	BBCWS-IE	World Business Report	2-6
1120	BBCWS-IE	World Business Report	2-6
1520	BBCWS-IE	World Business Report	2-6
1820	BBCWS-IE	World Business Report	2-6
0120	BBCWS-NX	World Business Report	2-6
0420	BBCWS-NX	World Business Report	2-6
0820	BBCWS-NX	World Business Report	2-6
1120	BBCWS-NX	World Business Report	2-6
1540	BBCWS-NX	World Business Report	2-6
1820	BBCWS-NX	World Business Report	2-6
1920	BBCWS-NX	World Business Report	2-6
2120	BBCWS-NX	World Business Report	2-6
2220	BBCWS-NX	World Business Report	1-5
0020	BBCWS-PR	World Business Report	7
0120	BBCWS-PR	World Business Report	2-6
0420	BBCWS-PR	World Business Report	2-7
0820	BBCWS-PR	World Business Report	2-6
1120	BBCWS-PR	World Business Report	2-6
1550	BBCWS-PR	World Business Report	2-6
1820	BBCWS-PR	World Business Report	2-6
1920	BBCWS-PR	World Business Report	2-6
2120	BBCWS-PR	World Business Report	2-6
2220	BBCWS-PR	World Business Report	1-5
0030	DW	World in Progress	3
0430	DW	World in Progress	2
0630	DW	World in Progress	2
0830	DW	World in Progress	2
0930	DW	World in Progress	1
1030	DW	World in Progress	2
1230	DW	World in Progress	2
1430	DW	World in Progress	2
1630	DW	World in Progress	2
1830	DW	World in Progress	2
2030	DW	World in Progress	2
2230	DW	World in Progress	2
0030	WRN-NA	World Vision	1
1500	WRN-NA	World Vision	7
1100	BBC-R4	You and Yours	2-6

Current Affairs Programs

Time	Station/Network	Program Name	Day(s)
0100	RNZ-NAT	Afternoons with Jim Mora	2-6
1300	BBC-R4	Any Answers?	7
1210	BBC-R4	Any Questions?	7
1900	BBC-R4	Any Questions?	6
0200	ABC-RA	Asia Review	7
0900	ABC-RA	Asia Review	7
1100	ABC-RA	Asia Review	7
1900	ABC-RA	Asia Review	6
2300	ABC-RA	Asia Review	6
0300	ABC-RN	Counterpoint	6
0600	ABC-RN	Counterpoint	2
1600	ABC-RN	Counterpoint	6
0000	ABC-RA	In the Loop	7

Current Affairs Programs *continued*

Time	Station/Network	Program Name	Day(s)
0100	ABC-RA	In the Loop	1
0310	ABC-RA	In the Loop	2-6
0400	ABC-RA	In the Loop	1
1730	ABC-RA	In the Loop	2-6
1200	ABC-RA	Late Night Live	2-6
0600	ABC-RN	Late Night Live	3-6
1200	ABC-RN	Late Night Live	2-6
1700	ABC-RA	Late Night Live Classic	7
1800	ABC-RN	Late Night Live Classic	7
0100	RTE-R1	Marian Finucane	1/2
1000	RTE-R1	Marian Finucane	1/7
1030	RTE-R1X	Marian Finucane (joined in progress)	1
0800	BBC-R4	Midweek	4
2030	BBC-R4	Midweek	4
1400	RTE-R1	Mooney	2-6
2100	RNZ-NAT	Nine to Noon with Kathryn Ryan	1-5
0530	ABC-RA	Pacific Beat - On the Mat	4-6
0535	ABC-RA	Pacific Beat - On the Mat	2/3
0735	ABC-RA	Pacific Beat - On the Mat	2-6
1300	WRN-NA	RTE Ireland: The Tubridy Show	2-6
2130	ABC-RA	Saturday Extra	6
2130	ABC-RN	Saturday Extra	6
0900	ABC-RN	Saturday Extra (from 2/09)	7
1500	ABC-RN	Saturday Extra - part 2	7
2030	BBC-R4	Start the Week	2
0130	ABC-RA	Talking Point	1
0430	ABC-RA	Talking Point	1
0615	ABC-RA	Talking Point	2/4-6
0615	ABC-RA	Talking Point	3
1020	ABC-RA	Talking Point	7
2330	ABC-RA	Talking Point	6
0000	BBCWS-AM	The Forum	2
0500	BBCWS-AM	The Forum	2
1400	BBCWS-AM	The Forum	1
0800	BBCWS-IE	The Forum	1
1300	BBCWS-IE	The Forum	1
1900	BBCWS-IE	The Forum	1
2300	BBCWS-IE	The Forum	1
1500	BBCWS-PR	The Forum	1
2100	BBCWS-PR	The Forum	1
1405	RNZ-NAT	The Forum	3
0200	ABC-RN	The National Interest	1
0800	ABC-RN	The National Interest	6
1700	ABC-RN	The National Interest	6
2300	CBC-R1A	The Night Time Review	3-6
0100	CBC-R1C	The Night Time Review	4-7
0000	CBC-R1E	The Night Time Review	4-7
0200	CBC-R1M	The Night Time Review	4-7
0300	CBC-R1P	The Night Time Review	4-7
1700	CBC-R1A	The Point	2-6
1900	CBC-R1C	The Point	2-6
1800	CBC-R1E	The Point	2-6
2000	CBC-R1M	The Point	2-6
2100	CBC-R1P	The Point	2-6
0600	CBC-R1S	The Point	3-7
1600	CBC-R1S	The Point	2-6
1200	CBC-R1A	The Sunday Edition	1
1400	CBC-R1C	The Sunday Edition	1
1300	CBC-R1E	The Sunday Edition	1
1500	CBC-R1M	The Sunday Edition	1
1600	CBC-R1P	The Sunday Edition	1
1113	CBC-R1S	The Sunday Edition	1
0130	RTE-R1	Today with Pat Kenny	3-7
0900	RTE-R1	Today with Pat Kenny	2-6

Society, Customs, Sites, Peoples & Cultural Values Programs

Time	Station/Network	Program Name	Day(s)
0750	BBC-R4	A Point of View	1
0000	RNW	Bridges with Africa	4
0030	RNW	Bridges with Africa	7
0100	RNW	Bridges with Africa	4
0130	RNW	Bridges with Africa	2/7
0200	RNW	Bridges with Africa	4
0230	RNW	Bridges with Africa	7
0330	RNW	Bridges with Africa	7
0405	RNW	Bridges with Africa	7
0430	RNW	Bridges with Africa	7
0500	RNW	Bridges with Africa	4
0505	RNW	Bridges with Africa	7
0530	RNW	Bridges with Africa	7
0600	RNW	Bridges with Africa	4
0605	RNW	Bridges with Africa	7
0700	RNW	Bridges with Africa	4
0705	RNW	Bridges with Africa	7
0730	RNW	Bridges with Africa	7
0800	RNW	Bridges with Africa	4
0805	RNW	Bridges with Africa	7
0830	RNW	Bridges with Africa	2/7
0900	RNW	Bridges with Africa	6
0930	RNW	Bridges with Africa	3
1000	RNW	Bridges with Africa	6
1030	RNW	Bridges with Africa	3
1100	RNW	Bridges with Africa	6
1130	RNW	Bridges with Africa	1/3
1200	RNW	Bridges with Africa	6
1230	RNW	Bridges with Africa	3
1330	RNW	Bridges with Africa	6
1405	RNW	Bridges with Africa	1/3
1530	RNW	Bridges with Africa	1/6
1630	RNW	Bridges with Africa	3
1730	RNW	Bridges with Africa	6
1805	RNW	Bridges with Africa	1
1830	RNW	Bridges with Africa	3
1930	RNW	Bridges with Africa	6
2000	RNW	Bridges with Africa	1
2030	RNW	Bridges with Africa	3
2130	RNW	Bridges with Africa	6
2230	RNW	Bridges with Africa	6
2300	RNW	Bridges with Africa	3
2330	RNW	Bridges with Africa	6
0900	SABC-SAFM	Church Service	1
0530	DW	Cool	5
0730	DW	Cool	5
0930	DW	Cool	5
1130	DW	Cool	5
1330	DW	Cool	5
1530	DW	Cool	5
1730	DW	Cool	5
1930	DW	Cool	5
2030	DW	Cool	1
2130	DW	Cool	5
2330	DW	Cool	5
0730	RFI	Crossroads	3
1430	RFI	Crossroads	3
0330	CBC-R1A	Culture Shock	3
0530	CBC-R1C	Culture Shock	3
0430	CBC-R1E	Culture Shock	3
0630	CBC-R1M	Culture Shock	3
0730	CBC-R1P	Culture Shock	3
0110	R.PRG	Czechs Today (monthly)	5
0210	R.PRG	Czechs Today (monthly)	5
0410	R.PRG	Czechs Today (monthly)	5
0440	R.PRG	Czechs Today (monthly)	5
0810	R.PRG	Czechs Today (monthly)	4
1140	R.PRG	Czechs Today (monthly)	4
1410	R.PRG	Czechs Today (monthly)	4
1710	R.PRG	Czechs Today (monthly)	4
1810	R.PRG	Czechs Today (monthly)	4
2110	R.PRG	Czechs Today (monthly)	4
2240	R.PRG	Czechs Today (monthly)	4

Time	Station/Network	Program Name	Day(s)
2340	R.PRG	Czechs Today (monthly)	4
1600	CBC-R1A	Definitely Not the Opera	7
1800	CBC-R1C	Definitely Not the Opera	7
1700	CBC-R1E	Definitely Not the Opera	7
1900	CBC-R1M	Definitely Not the Opera	7
2000	CBC-R1P	Definitely Not the Opera	7
0200	CBC-R1S	Definitely Not the Opera	1
1500	CBC-R1S	Definitely Not the Opera	7
0530	DW	Dialogue	6
0730	DW	Dialogue	6
0930	DW	Dialogue	6
1130	DW	Dialogue	6
1330	DW	Dialogue	6
1530	DW	Dialogue	6
1630	DW	Dialogue	7
1730	DW	Dialogue	6
1830	DW	Dialogue	7
1930	DW	Dialogue	6
2030	DW	Dialogue	7
2130	DW	Dialogue	6
2330	DW	Dialogue	6
0700	WRS-SUI	Dig It	1
0830	WRS-SUI	Dig It	5
0900	ABC-RN	Encounter	4
1800	ABC-RN	Encounter	4
2100	ABC-RN	Encounter	7
0245	KBS-WR	Faces of Korea	2
0345	KBS-WR	Faces of Korea	2
0515	KBS-WR	Faces of Korea	2
0645	KBS-WR	Faces of Korea	2
0815	KBS-WR	Faces of Korea	2
0915	KBS-WR	Faces of Korea	2
1145	KBS-WR	Faces of Korea	2
1215	KBS-WR	Faces of Korea	2
1245	KBS-WR	Faces of Korea	2
1515	KBS-WR	Faces of Korea	2
2045	KBS-WR	Faces of Korea	2
2245	KBS-WR	Faces of Korea	2
2345	KBS-WR	Faces of Korea	2
1700	SABC-SAFM	Faith 2 Faith	1
0600	BBCWS-AM	Heart and Soul	2
1900	BBCWS-AM	Heart and Soul	1
0530	BBCWS-IE	Heart and Soul	1
0030	BBCWS-PR	Heart and Soul	7
0730	BBCWS-PR	Heart and Soul	3
1830	BBCWS-PR	Heart and Soul	3
2230	BBCWS-PR	Heart and Soul	7
1535	RNZ-NAT	Hymns for Sunday Morning	7
1900	RNZ-NAT	Hymns for Sunday Morning	7
1940	BBC-R4	In Touch	3
1000	SABC-SAFM	Intune	7
0230	VOR-WS	Kaleidoscope	3/6
0530	VOR-WS	Kaleidoscope	1
0630	VOR-WS	Kaleidoscope	3
0730	VOR-WS	Kaleidoscope	2/6
0830	VOR-WS	Kaleidoscope	7
0930	VOR-WS	Kaleidoscope	3
0800	WRS-SUI	Kids in Mind	7
0830	WRS-SUI	Kids in Mind	6
1500	BBC-R4	Last Word	6
1930	BBC-R4	Last Word	1
0900	RTE-R1X	Mass	1
1600	ABC-RN	New Dimensions	1
1535	RNZ-NAT	New Zealand Society	1
0706	RNZ-NAT	One in Five	1
1230	RNZ-NAT	One in Five	5
0400	PR-EXT	Open Air	1
0415	PR-EXT	Open Air	5
0745	PR-EXT	Open Air	5
1730	PR-EXT	Open Air	7
2000	PR-EXT	Open Air	7
1100	SABC-SAFM	Otherwise	2-6
0110	R.PRG	Panorama	6

Time	Station/Network	Program Name	Day(s)
0210	R.PRG	Panorama	6
0410	R.PRG	Panorama	6
0440	R.PRG	Panorama	6
0810	R.PRG	Panorama	5
1140	R.PRG	Panorama	5
1410	R.PRG	Panorama	5
1710	R.PRG	Panorama	5
1810	R.PRG	Panorama	5
2110	R.PRG	Panorama	5
2240	R.PRG	Panorama	5
2340	R.PRG	Panorama	5
0443	BBC-R4	Prayer for the Day	2-7
1030	ABC-RA	Religion Report (til 1/09)	4
1530	ABC-RA	Religion Report (til 1/09)	4
1000	ABC-RN	Religion Report (til 1/09)	4
1530	ABC-RN	Religion Report (til 1/09)	4
2230	ABC-RN	Religion Report (til 1/09)	3
0730	RFI	Rendezvous	5
1430	RFI	Rendezvous	5
0430	BBCWS-AM	Reporting Religion	1
0830	BBCWS-AM	Reporting Religion	1
1130	BBCWS-AM	Reporting Religion	1
2330	BBCWS-AM	Reporting Religion	7
1130	BBCWS-IE	Reporting Religion	1
0130	BBCWS-NX	Reporting Religion	1
0430	BBCWS-NX	Reporting Religion	1
0630	BBCWS-NX	Reporting Religion	7
0830	BBCWS-NX	Reporting Religion	1
1030	BBCWS-NX	Reporting Religion	7
1330	BBCWS-NX	Reporting Religion	1
1930	BBCWS-NX	Reporting Religion	1
2330	BBCWS-NX	Reporting Religion	7
0030	BBCWS-PR	Reporting Religion	1
0330	BBCWS-PR	Reporting Religion	1
0830	BBCWS-PR	Reporting Religion	1
1330	BBCWS-PR	Reporting Religion	1
1930	BBCWS-PR	Reporting Religion	1
2330	BBCWS-PR	Reporting Religion	7
0800	BBC-R4	Saturday Live	7
0945	RTE-R1X	Service of the Word	1
0500	BBC-R4	Something Understood	1
2230	BBC-R4	Something Understood	1
1705	RTE-R1	Spirit Moves	1
0430	VOR-WS	Spiritual Flowerbed	4
0450	VOR-WS	Spiritual Flowerbed	3/5
0830	WRS-SUI	Stir It Up	3
0610	BBC-R4	Sunday	1
1200	ABC-RA	Sunday Night Talk	1
0710	BBC-R4	Sunday Worship	1
1700	CBC-R1A	Tapestry	1
1900	CBC-R1C	Tapestry	1
1800	CBC-R1E	Tapestry	1
2200	CBC-R1M	Tapestry	1
2200	CBC-R1P	Tapestry	1
0900	CBC-R1S	Tapestry	2
1400	CBC-R1S	Tapestry	1
1300	RTHK-3	Teen Time	2-6
1800	CBC-R1A	The Age of Persuasion	5
0130	CBC-R1S	The Age of Persuasion	1
1700	RTE-R1	The Angelus	1/7
0300	ABC-RA	The Ark (til 1/09)	1
0600	ABC-RA	The Ark (til 1/09)	1
0545	ABC-RN	The Ark (til 1/09)	1
1035	ABC-RN	The Ark (til 1/09)	4
0230	VOR-WS	The Christian Message from Moscow	6
0630	VOR-WS	The Christian Message from Moscow	7
0930	VOR-WS	The Christian Message from Moscow	7
1130	BBC-R4	The Food Programme	1
1500	BBC-R4	The Food Programme	2
0315	SABC-CHAF	The Inner Voice	3
0515	SABC-CHAF	The Inner Voice	1
0355	PR-EXT	The Kids	4
0405	PR-EXT	The Kids	3

Society, Customs, Sites, Peoples & Cultural Values Programs continued

Time	Station/Network	Program Name	Day(s)
0740	PR-EXT	The Kids	3
1725	PR-EXT	The Kids	1
1955	PR-EXT	The Kids	1
1930	BBC-R4	The Learning Curve	2
2200	BBC-R4	The Learning Curve	1
1900	BBC-R4	The Moral Maze	4
2115	BBC-R4	The Moral Maze	7
0445	BBC-R4	The Other Heartlands	1
0000	ABC-RA	The Spirit of Things	1
0300	ABC-RN	The Spirit of Things	3
0800	ABC-RN	The Spirit of Things	1
1600	ABC-RN	The Spirit of Things	3
0200	RNW	The State We're In	1
0400	RNW	The State We're In	1
0405	RNW	The State We're In	1
0500	RNW	The State We're In	1
0505	RNW	The State We're In	1
0605	RNW	The State We're In	1
0700	RNW	The State We're In	1
0705	RNW	The State We're In	1
0805	RNW	The State We're In	1
0900	RNW	The State We're In	7
1000	RNW	The State We're In	7
1200	RNW	The State We're In	7
1405	RNW	The State We're In	7
1605	RNW	The State We're In	7
1705	RNW	The State We're In	7
1905	RNW	The State We're In	7
2100	RNW	The State We're In	7
2200	RNW	The State We're In	7
0000	RNW	The State We're In	1
0030	RNW	The State We're In Midweek Edition	4
0130	RNW	The State We're In Midweek Edition	4
0230	RNW	The State We're In Midweek Edition	4
0330	RNW	The State We're In Midweek Edition	4
0405	RNW	The State We're In Midweek Edition	4
0430	RNW	The State We're In Midweek Edition	4
0505	RNW	The State We're In Midweek Edition	4
0530	RNW	The State We're In Midweek Edition	4
0605	RNW	The State We're In Midweek Edition	4
0630	RNW	The State We're In Midweek Edition	4
0705	RNW	The State We're In Midweek Edition	4
0730	RNW	The State We're In Midweek Edition	4
0805	RNW	The State We're In Midweek Edition	4
0830	RNW	The State We're In Midweek Edition	4
0900	RNW	The State We're In Midweek Edition	3
1000	RNW	The State We're In Midweek Edition	3
1100	RNW	The State We're In Midweek Edition	3
1200	RNW	The State We're In Midweek Edition	3
1330	RNW	The State We're In Midweek Edition	3
1530	RNW	The State We're In Midweek Edition	3
1730	RNW	The State We're In Midweek Edition	3
1930	RNW	The State We're In Midweek Edition	3
2130	RNW	The State We're In Midweek Edition	3
2230	RNW	The State We're In Midweek Edition	3
2330	RNW	The State We're In Midweek Edition	3
0030	BBCWS-AM	The Strand	3-7
1613	CBC-R1S	The Sunday Edition	1
0355	SABC-SAFM	This New Day	1/7
0025	RTHK-3	Thought for the Week	1
1406	RNZ-NAT	Touchstone	7
0730	CRI-WASH	Vogue	1-7
1300	CRI-WASH	Voices from Other Lands	5
1500	BBC-R4	Weekend Woman's Hour	7
0900	BBC-R4	Woman's Hour	2-6
0210	RTI	Women Making Waves	2

Everyday Domestic Life Programs

Time	Station/Network	Program Name	Day(s)
1750	RNZ-NAT	Auckland Stories	7
1830	ABC-RA	Australia All Over	7
0900	ABC-RA	Australia Talks	2-6
1600	ABC-RA	Australia Talks	2-5
0800	ABC-RN	Australia Talks	3-5
1700	ABC-RN	Australia Talks	2-5
0800	ABC-RN	Australia Talks (til 1/09)	2
0030	ABC-RA	Australian Bite	7
0400	ABC-RA	Australian Bite	7
0500	ABC-RA	Australian Bite	7
1330	ABC-RA	Australian Bite	3
1700	ABC-RA	Australian Bite	3
1830	ABC-RA	Australian Bite	6
1400	ABC-RA	AWAYE!	3
0500	ABC-RN	AWAYE!	2
0800	ABC-RN	AWAYE!	7
1700	ABC-RN	AWAYE!	7
1800	ABC-RN	AWAYE!	1
0030	RTHK-3	Backchat	2-6
0530	WRN-NA	Banns Radio International: Copenhagen Calling	7
0100	ABC-RN	Bush Telegraph	2-6
1500	ABC-RN	Bush Telegraph (from 2/09)	2-6
1400	CBC-R1A	C'est la vie	3
2330	CBC-R1A	C'est la vie	1
0030	CBC-R1C	C'est la vie	2
1600	CBC-R1C	C'est la vie	3
1500	CBC-R1E	C'est la vie	3
2330	CBC-R1E	C'est la vie	1
0130	CBC-R1M	C'est la vie	2
1700	CBC-R1M	C'est la vie	3
0230	CBC-R1P	C'est la vie	2
1800	CBC-R1P	C'est la vie	3
2030	CBC-R1S	C'est la vie	3
2230	CBC-R1S	C'est la vie	1
1530	RNZ-NAT	Canterbury Tales	5
0027	CRI-ENG	China Horizons	4
0127	CRI-ENG	China Horizons	4
0327	CRI-ENG	China Horizons	4
0427	CRI-ENG	China Horizons	4
0527	CRI-ENG	China Horizons	4
0627	CRI-ENG	China Horizons	4
1327	CRI-ENG	China Horizons	3
1427	CRI-ENG	China Horizons	3
1527	CRI-ENG	China Horizons	3
0030	CRI-RTC	China Horizons	1
0130	CRI-RTC	China Horizons	1
0230	CRI-RTC	China Horizons	1
0330	CRI-RTC	China Horizons	1
0430	CRI-RTC	China Horizons	1
0530	CRI-RTC	China Horizons	1
0630	CRI-RTC	China Horizons	1
0830	CRI-RTC	China Horizons	1
0930	CRI-RTC	China Horizons	1
1230	CRI-RTC	China Horizons	1
1330	CRI-RTC	China Horizons	1
1430	CRI-RTC	China Horizons	1
1530	CRI-RTC	China Horizons	1
1630	CRI-RTC	China Horizons	1
1830	CRI-RTC	China Horizons	1
1930	CRI-RTC	China Horizons	1
2030	CRI-RTC	China Horizons	1
2130	CRI-RTC	China Horizons	1
2330	CRI-RTC	China Horizons	7
0220	CRI-WASH	China Horizons	4-1
1300	CRI-WASH	China Horizons	7
1330	CRI-WASH	China Horizons	1
1420	CRI-WASH	China Horizons	3-7
2000	CRI-WASH	China Horizons	2
2000	ABC-RN	Country Breakfast	6
0904	RNZ-NAT	Country Life	6
1900	RNZ-NAT	Country Life	6
2000	CBC-R1A	Cross Country Checkup	1
2000	CBC-R1C	Cross Country Checkup	1
2000	CBC-R1E	Cross Country Checkup	1
2000	CBC-R1M	Cross Country Checkup	1

Time	Station/Network	Program Name	Day(s)
2000	CBC-R1P	Cross Country Checkup	1
2000	CBC-R1S	Cross Country Checkup	1
0000	RNW	Curious Orange	5
0030	RNW	Curious Orange	3
0100	RNW	Curious Orange	5
0130	RNW	Curious Orange	1/3
0200	RNW	Curious Orange	5
0230	RNW	Curious Orange	3
0330	RNW	Curious Orange	1
0330	RNW	Curious Orange	3
0405	RNW	Curious Orange	3
0430	RNW	Curious Orange	3
0500	RNW	Curious Orange	5
0505	RNW	Curious Orange	3
0530	RNW	Curious Orange	3
0600	RNW	Curious Orange	5
0605	RNW	Curious Orange	3
0630	RNW	Curious Orange	1/3
0700	RNW	Curious Orange	5
0705	RNW	Curious Orange	3
0730	RNW	Curious Orange	3
0800	RNW	Curious Orange	5
0805	RNW	Curious Orange	3
0830	RNW	Curious Orange	1/3
0900	RNW	Curious Orange	2
0930	RNW	Curious Orange	4
1000	RNW	Curious Orange	2
1030	RNW	Curious Orange	4
1100	RNW	Curious Orange	2
1130	RNW	Curious Orange	4/7
1200	RNW	Curious Orange	2
1230	RNW	Curious Orange	4
1330	RNW	Curious Orange	2
1330	RNW	Curious Orange	7
1400	RNW	Curious Orange	4
1530	RNW	Curious Orange	2
1630	RNW	Curious Orange	4
1730	RNW	Curious Orange	2
1830	RNW	Curious Orange	4/7
1930	RNW	Curious Orange	2
2030	RNW	Curious Orange	4/7
2130	RNW	Curious Orange	2
2130	RNW	Curious Orange	7
2230	RNW	Curious Orange	2
2300	RNW	Curious Orange	4
2330	RNW	Curious Orange	2/7
1300	BBC-R4	Gardeners' Question Time	1
1400	BBC-R4	Gardeners' Question Time	4
1710	RNZ-NAT	He Rourou (in Maori)	1-5
0210	RTI	Ilha Formosa	5
0027	CRI-ENG	In the Spotlight	1
0127	CRI-ENG	In the Spotlight	1
0327	CRI-ENG	In the Spotlight	1
0427	CRI-ENG	In the Spotlight	1
0527	CRI-ENG	In the Spotlight	1
0627	CRI-ENG	In the Spotlight	1
1327	CRI-ENG	In the Spotlight	7
1427	CRI-ENG	In the Spotlight	7
1527	CRI-ENG	In the Spotlight	7
0030	CRI-RTC	In the Spotlight	4
0130	CRI-RTC	In the Spotlight	4
0230	CRI-RTC	In the Spotlight	4
0330	CRI-RTC	In the Spotlight	4
0430	CRI-RTC	In the Spotlight	4
0530	CRI-RTC	In the Spotlight	4
0630	CRI-RTC	In the Spotlight	4
0830	CRI-RTC	In the Spotlight	4
0930	CRI-RTC	In the Spotlight	4
1230	CRI-RTC	In the Spotlight	4
1330	CRI-RTC	In the Spotlight	4
1430	CRI-RTC	In the Spotlight	4
1530	CRI-RTC	In the Spotlight	4
1630	CRI-RTC	In the Spotlight	4
1830	CRI-RTC	In the Spotlight	4
1930	CRI-RTC	In the Spotlight	4

Time	Station/Network	Program Name	Day(s)
2030	CRI-RTC	In the Spotlight	4
2130	CRI-RTC	In the Spotlight	4
1300	CRI-WASH	In the Spotlight	4
1330	CRI-WASH	In the Spotlight	5
2000	CRI-WASH	In the Spotlight	5
0245	KBS-WR	Korea Today and Tomorrow	5
0345	KBS-WR	Korea Today and Tomorrow	5
0515	KBS-WR	Korea Today and Tomorrow	5
0645	KBS-WR	Korea Today and Tomorrow	5
0815	KBS-WR	Korea Today and Tomorrow	5
0915	KBS-WR	Korea Today and Tomorrow	5
1145	KBS-WR	Korea Today and Tomorrow	5
1215	KBS-WR	Korea Today and Tomorrow	5
1245	KBS-WR	Korea Today and Tomorrow	5
1515	KBS-WR	Korea Today and Tomorrow	5
2045	KBS-WR	Korea Today and Tomorrow	5
2245	KBS-WR	Korea Today and Tomorrow	5
2345	KBS-WR	Korea Today and Tomorrow	5
0330	PR-EXT	Letter from Poland	5
0705	PR-EXT	Letter from Poland	5
0725	PR-EXT	Letter from Poland	6
1720	PR-EXT	Letter from Poland	3
1950	PR-EXT	Letter from Poland	3
0115	R.PRG	Letter from Prague	2
0215	R.PRG	Letter from Prague	2
0415	R.PRG	Letter from Prague	2
0445	R.PRG	Letter from Prague	2
0815	R.PRG	Letter from Prague	1
1145	R.PRG	Letter from Prague	1
1415	R.PRG	Letter from Prague	1
1715	R.PRG	Letter from Prague	1
1815	R.PRG	Letter from Prague	1
2115	R.PRG	Letter from Prague	1
2245	R.PRG	Letter from Prague	1
2345	R.PRG	Letter from Prague	1
0027	CRI-ENG	Life in China	6
0127	CRI-ENG	Life in China	6
0327	CRI-ENG	Life in China	6
0427	CRI-ENG	Life in China	6
0527	CRI-ENG	Life in China	6
0627	CRI-ENG	Life in China	6
1327	CRI-ENG	Life in China	5
1427	CRI-ENG	Life in China	5
1527	CRI-ENG	Life in China	5
0030	CRI-RTC	Life in China	6
0130	CRI-RTC	Life in China	6
0230	CRI-RTC	Life in China	6
0330	CRI-RTC	Life in China	6
0430	CRI-RTC	Life in China	6
0530	CRI-RTC	Life in China	6
0630	CRI-RTC	Life in China	6
0830	CRI-RTC	Life in China	6
0930	CRI-RTC	Life in China	6
1230	CRI-RTC	Life in China	6
1330	CRI-RTC	Life in China	6
1430	CRI-RTC	Life in China	6
1530	CRI-RTC	Life in China	6
1630	CRI-RTC	Life in China	6
1830	CRI-RTC	Life in China	6
1930	CRI-RTC	Life in China	6
2030	CRI-RTC	Life in China	6
2130	CRI-RTC	Life in China	6
1300	CRI-WASH	Life in China	1/6
1330	CRI-WASH	Life in China	7
1100	ABC-RN	Life Matters	2-5
2300	ABC-RN	Life Matters	1-5
0105	R.PRG	Magazine	1
0205	R.PRG	Magazine	1
0405	R.PRG	Magazine	1
0435	R.PRG	Magazine	1
0805	R.PRG	Magazine	7
1135	R.PRG	Magazine	7
1405	R.PRG	Magazine	7
1705	R.PRG	Magazine	7
1805	R.PRG	Magazine	7

Time	Station/Network	Program Name	Day(s)
2105	R.PRG	Magazine	7
2235	R.PRG	Magazine	7
2335	R.PRG	Magazine	7
2355	CRI-RTC	Music	1-7
1835	RNZI	News about New Zealand	2-6
1710	RNZ-NAT	Nga Marae	7
0535	BBC-R4	On Your Farm	1
0507	BBC-R4	Open Country	7
1230	BBC-R4	Open Country	5
2345	CBC-R1A	Outfront	3-6
0145	CBC-R1C	Outfront	4-7
0045	CBC-R1E	Outfront	4-7
0245	CBC-R1M	Outfront	4-7
0345	CBC-R1P	Outfront	4-7
0745	CBC-R1S	Outfront	3-7
1745	CBC-R1S	Outfront	2-6
0220	RTI	People	4
0030	WRN-NA	Radio Guangdong: Guangdong Today	2
0830	WRN-NA	Radio Guangdong: Guangdong Today	7
1600	WRN-NA	Radio Guangdong: Guangdong Today	7
1400	BBC-R4	Ramblings	6
1330	ABC-RA	Rural Reporter	4
1700	ABC-RA	Rural Reporter	4
1930	ABC-RA	Rural Reporter	6
0700	SABC-SAFM	SAfm Lifestyle	7
0245	KBS-WR	Seoul Report	6
0345	KBS-WR	Seoul Report	6
0515	KBS-WR	Seoul Report	6
0645	KBS-WR	Seoul Report	6
0815	KBS-WR	Seoul Report	6
0915	KBS-WR	Seoul Report	6
1145	KBS-WR	Seoul Report	6
1215	KBS-WR	Seoul Report	6
1245	KBS-WR	Seoul Report	6
1515	KBS-WR	Seoul Report	6
2045	KBS-WR	Seoul Report	6
2245	KBS-WR	Seoul Report	6
2245	KBS-WR	Seoul Report	6
2345	KBS-WR	Seoul Report	6
1130	ABC-RA	Speaking Out	1
0000	RNZ-NAT	Spectrum	1
0730	RNZ-NAT	Spectrum	5
1230	RNZ-NAT	Spectrum	4
1705	RTE-R1	Spectrum	7
0330	RNZI	Spectrum (fortnightly)	3
0730	RNZI	Spectrum (fortnightly)	3
1130	RNZI	Spectrum (fortnightly)	2
1330	RNZI	Spectrum (fortnightly)	2
1630	RNZI	Spectrum (fortnightly)	2
2035	RNZI	Spectrum (fortnightly)	6
0220	RTI	Spotlight	1
0330	ABC-RN	Street Stories (til 1/09)	1
1000	ABC-RN	Street Stories (til 1/09)	1
0431	WRS-SUI	Switzerland Today	2-6
0530	WRS-SUI	Switzerland Today	2-6
1030	WRS-SUI	Switzerland Today	2-6
1130	WRS-SUI	Switzerland Today	2-6
1431	WRS-SUI	Switzerland Today	2-6
1530	WRS-SUI	Switzerland Today	2-6
0530	RNZ-NAT	Tagata o te Moana	7
0500	RNZI	Tagata o te Moana	7
1300	RNZI	Tagata o te Moana	1/7
1815	RNZI	Tagata o te Moana	1
1910	RNZI	Tagata o te Moana	6
0110	R.PRG	Talking Point	4
0210	R.PRG	Talking Point	4
0410	R.PRG	Talking Point	4
0810	R.PRG	Talking Point	3
1140	R.PRG	Talking Point	3
1410	R.PRG	Talking Point	3
1710	R.PRG	Talking Point	3
1810	R.PRG	Talking Point	3
2110	R.PRG	Talking Point	3
2240	R.PRG	Talking Point	3
2340	R.PRG	Talking Point	3

Everyday Domestic Life Programs *continued*

Time	Station/Network	Program Name	Day(s)
0730	RNZ-NAT	Te Ahi Kaa	1
1315	RNZ-NAT	Te Ahi Kaa	1
0700	WRS-SUI	The Classifieds	6
0655	BBC-R4	The Radio 4 Appeal	1
1427	BBC-R4	The Radio 4 Appeal	5
2026	BBC-R4	The Radio 4 Appeal	1
0400	VOR-WS	This is Russia	2
0500	VOR-WS	This is Russia	7
0600	VOR-WS	This is Russia	1
0700	VOR-WS	This is Russia	3/6
0800	VOR-WS	This is Russia	4
0900	VOR-WS	This is Russia	1/2
0230	VOR-WS	Timelines	2
0530	VOR-WS	Timelines	7
0730	VOR-WS	Timelines	1
0930	VOR-WS	Timelines	1
0425	RFI	Today in France	2-6
0525	RFI	Today in France	2-6
0725	RFI	Today in France	2-6
1420	RFI	Today in France	2-7
0027	CRI-ENG	Voices from Other Lands	5
0127	CRI-ENG	Voices from Other Lands	5
0327	CRI-ENG	Voices from Other Lands	5
0427	CRI-ENG	Voices from Other Lands	5
0527	CRI-ENG	Voices from Other Lands	5
0627	CRI-ENG	Voices from Other Lands	5
1327	CRI-ENG	Voices from Other Lands	4
1427	CRI-ENG	Voices from Other Lands	4
1527	CRI-ENG	Voices from Other Lands	4
0030	CRI-RTC	Voices from Other Lands	5
0130	CRI-RTC	Voices from Other Lands	5
0230	CRI-RTC	Voices from Other Lands	5
0330	CRI-RTC	Voices from Other Lands	5
0430	CRI-RTC	Voices from Other Lands	5
0530	CRI-RTC	Voices from Other Lands	5
0630	CRI-RTC	Voices from Other Lands	5
0830	CRI-RTC	Voices from Other Lands	5
0930	CRI-RTC	Voices from Other Lands	5
1230	CRI-RTC	Voices from Other Lands	5
1330	CRI-RTC	Voices from Other Lands	5
1430	CRI-RTC	Voices from Other Lands	5
1530	CRI-RTC	Voices from Other Lands	5
1630	CRI-RTC	Voices from Other Lands	5
1830	CRI-RTC	Voices from Other Lands	5
1930	CRI-RTC	Voices from Other Lands	5
2030	CRI-RTC	Voices from Other Lands	5
2130	CRI-RTC	Voices from Other Lands	5
1330	CRI-WASH	Voices from Other Lands	6
2000	CRI-WASH	Voices from Other Lands	6
0700	WRS-SUI	Your Space	2/4

Media & Communications Programs

Time	Station/Network	Program Name	Day(s)
0330	BBCWS-AM	Digital Planet	4
0830	BBCWS-AM	Digital Planet	5
1330	BBCWS-AM	Digital Planet	3
1430	BBCWS-AM	Digital Planet	3
2230	BBCWS-AM	Digital Planet	3
1230	BBCWS-IE	Digital Planet	3
1530	BBCWS-IE	Digital Planet	3
1930	BBCWS-IE	Digital Planet	3
2330	BBCWS-IE	Digital Planet	3
1735	RNZ-NAT	Digital Planet	6
0830	WRS-SUI	Gadget Guru	4
0900	WRS-SUI	Gadget Guru	1
0830	WRN-NA	Glenn Hauser's World of Radio	1
1730	WRN-NA	Glenn Hauser's World of Radio	1/7
0330	RNZI	Mailbox (fortnightly)	3
0730	RNZI	Mailbox (fortnightly)	3
1130	RNZI	Mailbox (fortnightly)	2
1330	RNZI	Mailbox (fortnightly)	2
1630	RNZI	Mailbox (fortnightly)	2

Media & Communications Programs *continued*

Time	Station/Network	Program Name	Day(s)
2035	RNZI	Mailbox (fortnightly)	6
1030	ABC-RA	Media Report (til 1/09)	5
1530	ABC-RA	Media Report (til 1/09)	5
1000	ABC-RN	Media Report (til 1/09)	5
1530	ABC-RN	Media Report (til 1/09)	5
2230	ABC-RN	Media Report (til 1/09)	4
0700	SABC-SAFM	Media@SAfm	1
1000	RNZ-NAT	Mediawatch	1
2110	RNZ-NAT	Mediawatch	7
0335	PR-EXT	Multimedia	6
0705	PR-EXT	Multimedia	6
1735	PR-EXT	Multimedia	4
2005	PR-EXT	Multimedia	4
1400	CBC-R1A	Spark	4
1900	CBC-R1A	Spark	7
1600	CBC-R1C	Spark	4
2100	CBC-R1C	Spark	7
1500	CBC-R1E	Spark	4
2000	CBC-R1E	Spark	7
1700	CBC-R1M	Spark	4
2200	CBC-R1M	Spark	7
1800	CBC-R1P	Spark	4
2300	CBC-R1P	Spark	7
1800	CBC-R1S	Spark	7
2030	CBC-R1S	Spark	4
2000	CBC-R1C	The Age of Persuasion	5
1900	CBC-R1E	The Age of Persuasion	5
2100	CBC-R1M	The Age of Persuasion	5
2200	CBC-R1P	The Age of Persuasion	5
1230	BBC-R4	The Media Show	4

Ideas, Philosophy & Learning Programs

Time	Station/Network	Program Name	Day(s)
1400	ABC-RA	Big Ideas	2
1700	ABC-RA	Big Ideas	6
0700	ABC-RN	Big Ideas	1
0900	ABC-RN	Big Ideas (til 1/09)	7
0220	RTI	Breakfast Club	5
0000	CBC-R1A	Ideas	3-7
0200	CBC-R1C	Ideas	3-7
0100	CBC-R1E	Ideas	3-7
0300	CBC-R1M	Ideas	3-7
0400	CBC-R1P	Ideas	3-7
0400	CBC-R1S	Ideas	3-7
2300	CBC-R1S	Ideas	2-6
1305	RNZ-NAT	Ideas	5
2305	RNZ-NAT	Ideas	7
0800	BBC-R4	In Our Time	5
2030	BBC-R4	In Our Time	5
0445	BBC-R4	iPM	7
1400	BBC-R4	Questions, Questions	5
1430	ABC-RA	The Philosopher's Zone	4
0335	ABC-RN	The Philosopher's Zone	2/7
1635	ABC-RN	The Philosopher's Zone	2
1500	BBC-R4	Thinking Allowed	4
2315	BBC-R4	Thinking Allowed	1=

Environment Programs

Time	Station/Network	Program Name	Day(s)
0000	RNW	Earthbeat	3
0030	RNW	Earthbeat	6
0100	RNW	Earthbeat	1/3
0130	RNW	Earthbeat	6
0200	RNW	Earthbeat	3
0230	RNW	Earthbeat	6
0330	RNW	Earthbeat	6
0405	RNW	Earthbeat	6
0430	RNW	Earthbeat	6
0500	RNW	Earthbeat	3

Environment Programs *continued*

Time	Station/Network	Program Name	Day(s)
0505	RNW	Earthbeat	6
0530	RNW	Earthbeat	6
0600	RNW	Earthbeat	1/3
0605	RNW	Earthbeat	6
0630	RNW	Earthbeat	6
0700	RNW	Earthbeat	3
0705	RNW	Earthbeat	6
0730	RNW	Earthbeat	6
0800	RNW	Earthbeat	1/3
0805	RNW	Earthbeat	6
0830	RNW	Earthbeat	6
0900	RNW	Earthbeat	5
0930	RNW	Earthbeat	2
1000	RNW	Earthbeat	5
1030	RNW	Earthbeat	2
1100	RNW	Earthbeat	5/7
1130	RNW	Earthbeat	2
1200	RNW	Earthbeat	5
1230	RNW	Earthbeat	2
1330	RNW	Earthbeat	5
1400	RNW	Earthbeat	2
1530	RNW	Earthbeat	5/7
1630	RNW	Earthbeat	2
1730	RNW	Earthbeat	5
1830	RNW	Earthbeat	2
1930	RNW	Earthbeat	5
2000	RNW	Earthbeat	7
2030	RNW	Earthbeat	2
2130	RNW	Earthbeat	5
2230	RNW	Earthbeat	5
2300	RNW	Earthbeat	2/7
2330	RNW	Earthbeat	5
0030	DW	Living Planet	6
0430	DW	Living Planet	5
0630	DW	Living Planet	5
0830	DW	Living Planet	5
1030	DW	Living Planet	5
1230	DW	Living Planet	5
1430	DW	Living Planet	5
1630	DW	Living Planet	5
1830	DW	Living Planet	5
2030	DW	Living Planet	5
2230	DW	Living Planet	5
0330	BBCWS-AM	One Planet	6
0630	BBCWS-AM	One Planet	2
0830	BBCWS-AM	One Planet	6
1030	BBCWS-AM	One Planet	1
1330	BBCWS-AM	One Planet	5
1430	BBCWS-AM	One Planet	5
1530	BBCWS-AM	One Planet	1
1930	BBCWS-AM	One Planet	1
2230	BBCWS-AM	One Planet	5
0530	BBCWS-IE	One Planet	7
1230	BBCWS-IE	One Planet	5
1530	BBCWS-IE	One Planet	5
1930	BBCWS-IE	One Planet	5
2330	BBCWS-IE	One Planet	5
0230	BBCWS-PR	One Planet	7
0730	BBCWS-PR	One Planet	5
1030	BBCWS-PR	One Planet	7
1830	BBCWS-PR	One Planet	5
2230	BBCWS-PR	One Planet	6
1315	RNZ-NAT	One Planet	2
0000	RNZ-NAT	This Way Up w/ Simon Morton	7

General Documentary Programs

Time	Station/Network	Program Name	Day(s)
0000	CBC-R1S	And The Winner Is	3
0800	CBC-R1S	And The Winner Is	3
1000	CBC-R1S	And The Winner Is	2
0300	CBC-R1A	And the Winner Is	5
0500	CBC-R1C	And the Winner Is	5
0400	CBC-R1E	And the Winner Is	5
0600	CBC-R1M	And the Winner Is	5
0700	CBC-R1P	And the Winner Is	5
0600	RTE-R1	Documentary series	7
0930	RTE-R1	Documentary series	1
1900	RTE-R1	Documentary series	1
0904	RNZ-NAT	The Tuesday Feature	3

General Interest Programs

Time	Station/Network	Program Name	Day(s)
0515	SABC-CHAF	37 Degrees	7
1305	RNZ-NAT	BBC Feature	4
1315	RNZ-NAT	BBC Feature	3
1405	RNZ-NAT	BBC Feature	2
0600	WRS-SUI	Best of the Week	7
0710	RTE-R1	Bowman	1
0400	CBC-R1A	CBC Overnight	1/7
0400	CBC-R1A	CBC Overnight	2-6
0600	CBC-R1C	CBC Overnight	1/7
0600	CBC-R1C	CBC Overnight	2-6
0500	CBC-R1E	CBC Overnight	1/7
0500	CBC-R1E	CBC Overnight	2-6
0700	CBC-R1M	CBC Overnight	1/7
0700	CBC-R1M	CBC Overnight	2-6
0800	CBC-R1P	CBC Overnight	1/7
0800	CBC-R1P	CBC Overnight	2-6
0400	RTE-R1	Conversations with Eamon Dunphy	1
1315	RNZ-NAT	From the World	6
0800	BBC-R4	General feature or documentary series	3
0930	BBC-R4	General feature or documentary series	7
1000	BBC-R4	General feature or documentary series	2
2030	BBC-R4	General feature or documentary series	3
2200	BBC-R4	General feature or documentary series	2
0600	RNZ-NAT	Great Encounters	7
1100	RTHK-3	Half hour programs in Nepalese and Urdu	1
0045	DW	Inspired Minds	1
0345	DW	Inspired Minds	1
0445	DW	Inspired Minds	1
0645	DW	Inspired Minds	1
0900	RTE-R1	Playback	7
2000	CBC-R1A	Regional performance	7
2200	CBC-R1C	Regional performance	7
2100	CBC-R1E	Regional performance	7
2300	CBC-R1M	Regional performance	7
0000	CBC-R1P	Regional performance	1
0300	CBC-R1A	Rewind	6
0500	CBC-R1C	Rewind	6
0400	CBC-R1E	Rewind	6
0600	CBC-R1M	Rewind	6
0700	CBC-R1P	Rewind	6
0000	CBC-R1S	Rewind	2
0500	CBC-R1S	Rewind	2-7
0600	CBC-R1S	Rewind	1
0900	CBC-R1S	Rewind	3-7
1800	CBC-R1S	Rewind	2-6
2040	RNZI	RNZI Feature	3-5
2140	RNZI	RNZI Feature	1-5
0500	CBC-R1S	Special Delivery	1
0600	CBC-R1S	Special Delivery	2
0900	CBC-R1S	Special Delivery	1
2015	SABC-CHAF	Talk or Debate	1-7
2115	SABC-CHAF	Talk or Debate	2-7
1900	BBC-R4	The Archive Hour	7
0300	CBC-R1A	The Choice	4
1800	CBC-R1A	The Choice	3

General Interest Programs continued

Time	Station/Network	Program Name	Day(s)
0500	CBC-R1C	The Choice	4
2000	CBC-R1C	The Choice	3
0400	CBC-R1E	The Choice	4
1900	CBC-R1E	The Choice	3
0600	CBC-R1M	The Choice	4
2100	CBC-R1M	The Choice	3
0700	CBC-R1P	The Choice	4
2200	CBC-R1P	The Choice	3
0000	CBC-R1S	The Choice	4
0230	CBC-R1S	The Choice	2
0800	CBC-R1S	The Choice	4
1000	CBC-R1S	The Choice	3
2230	CBC-R1S	The Choice	2
0407	RNZ-NAT	The Sunday Feature	1
0030	RTHK-3	The Week on Three	7
2200	RNZI	World and Pacific News	1-5

Government, Politics & The Law Programs

Time	Station/Network	Program Name	Day(s)
2040	RNZI	Focus on Politics	1
0030	CRI-RTC	Frontline	2
0130	CRI-RTC	Frontline	2
0230	CRI-RTC	Frontline	2
0330	CRI-RTC	Frontline	2
0430	CRI-RTC	Frontline	2
0530	CRI-RTC	Frontline	2
0630	CRI-RTC	Frontline	2
0830	CRI-RTC	Frontline	2
0930	CRI-RTC	Frontline	2
1230	CRI-RTC	Frontline	2
1330	CRI-RTC	Frontline	2
1430	CRI-RTC	Frontline	2
1530	CRI-RTC	Frontline	2
1630	CRI-RTC	Frontline	2
1830	CRI-RTC	Frontline	2
1930	CRI-RTC	Frontline	2
2030	CRI-RTC	Frontline	2
2130	CRI-RTC	Frontline	2
1300	CRI-WASH	Frontline	2
1330	CRI-WASH	Frontline	3
2000	CRI-WASH	Frontline	3
1500	BBC-R4	Law in Action	3
0630	ABC-RA	Law Report	3
1030	ABC-RA	Law Report	3
1530	ABC-RA	Law Report	3
2230	ABC-RN	Law Report	2
1900	SABC-SAFM	Law Report	2
1400	ABC-RN	Law Report (from 2/09)	3
1000	ABC-RN	Law Report (til 1/09)	3
1530	ABC-RN	Law Report (til 1/09)	3
0015	RTHK-3	Letter to Hong Kong	1
0027	CRI-ENG	People in the Know	2
0127	CRI-ENG	People in the Know	2
0327	CRI-ENG	People in the Know	2
0427	CRI-ENG	People in the Know	2
0527	CRI-ENG	People in the Know	2
0627	CRI-ENG	People in the Know	2
1327	CRI-ENG	People in the Know	1
1427	CRI-ENG	People in the Know	1
1527	CRI-ENG	People in the Know	1
2330	CRI-RTC	People in the Know	1-5
0200	CRI-WASH	People in the Know	4-1
1215	CRI-WASH	People in the Know	1-7
1400	CRI-WASH	People in the Know	3-7
2000	CRI-WASH	People in the Know	1/7
0230	BBCWS-AM	Politics UK	7
0930	BBCWS-AM	Politics UK	1
1830	BBCWS-AM	Politics UK	7
0030	BBCWS-IE	Politics UK	1

Government, Politics & The Law Programs *continued*

Time	Station/Network	Program Name	Day(s)
0430	BBCWS-IE	Politics UK	7
0930	BBCWS-IE	Politics UK	1
0030	BBCWS-NX	Politics UK	7
0230	BBCWS-NX	Politics UK	7
0330	BBCWS-NX	Politics UK	1
0930	BBCWS-NX	Politics UK	1
1130	BBCWS-NX	Politics UK	7
1430	BBCWS-NX	Politics UK	1
1830	BBCWS-NX	Politics UK	7
2130	BBCWS-NX	Politics UK	1
2230	BBCWS-NX	Politics UK	6/7
0930	BBCWS-PR	Politics UK	7
1430	BBCWS-PR	Politics UK	1
1830	BBCWS-PR	Politics UK	7
1300	RTE-R1	Saturday Sport with John Kenny	7
0200	RTE-R1	Saturdayview	1
1200	RTE-R1	Saturdayview	7
0315	SABC-CHAF	Tam Tam Express	4
0715	SABC-CHAF	Tam Tam Express	4
1200	CBC-R1A	The House	7
1400	CBC-R1C	The House	7
1300	CBC-R1E	The House	7
1500	CBC-R1M	The House	7
1600	CBC-R1P	The House	7
0400	CBC-R1S	The House	1
1113	CBC-R1S	The House	7
2200	RTE-R1	The Late Debate	2-5
1000	BBC-R4	The Week in Parliament	7
2100	BBC-R4	The Westminster Hour	1
2230	BBC-R4	Today in Parliament	2-6

General Magazine-Style Programs

Time	Station/Network	Program Name	Day(s)
0400	SABC-SAFM	AM Live	2-6
2300	CRI-WASH	Beyond Beijing	1-7
0430	WRS-SUI	Breakfast Show	2-6
1100	CRI-ENG	China Drive	2-6
2300	CRI-ENG	China Drive	1-5
0700	CRI-RTC	China Drive	2-6
1000	CRI-RTC	China Drive	2-6
1100	CRI-RTC	China Drive	2-6
1700	CRI-RTC	China Drive	2-6
2200	CRI-RTC	China Drive	2-6
0400	RNZ-NAT	Four 'til Eight with Liz Barry	1
0200	SABC-SAFM	Heads Up	2-6
2040	CRI-WASH	Listener's Garden	1-7
0900	CBC-R1A	Local morning program	1/7
1100	CBC-R1C	Local morning program	1/7
1000	CBC-R1E	Local morning program	1/7
1200	CBC-R1M	Local morning program	1/7
1300	CBC-R1P	Local morning program	1/7
1000	CBC-RCI	Masala Canada	1
1500	CBC-RCI	Masala Canada	7
1505	CBC-RCI	Masala Canada	7
2200	CBC-RCI	Masala Canada	7
2300	CBC-RCI	Masala Canada	7
2305	CBC-RCI	Masala Canada	7
0130	RTHK-3	Morning Brew	2-6
0515	RTHK-3	Naked Lunch	2-6
1010	RTHK-3	Neil Chase in New York	7
0800	WRS-SUI	On the Beat	2-6
0300	BBCWS-AM	Outlook	3-7
0630	BBCWS-AM	Outlook	3-7
1300	BBCWS-AM	Outlook	2-6
1830	BBCWS-AM	Outlook	2-6
2200	BBCWS-AM	Outlook	2-6
0230	BBCWS-IE	Outlook	3-7
0830	BBCWS-IE	Outlook	2-6

General Magazine-Style Programs *continued*

Time	Station/Network	Program Name	Day(s)
1400	BBCWS-IE	Outlook	2-6
2200	BBCWS-IE	Outlook	2-6
2130	BBCWS-PR	Outlook	2-6
0100	RTHK-3	Sunday Morning	1
2000	RNZ-NAT	Sunday Morning with Chris Laidlaw	7
0800	RTHK-3	Sunday PM	1
1500	CBC-RCI	The Link	2-6
1505	CBC-RCI	The Link	2-6
2300	CBC-RCI	The Link	2-6
2305	CBC-RCI	The Link (first hour)	2-6
0005	CBC-RCI	The Link (second hour)	3-7
1000	CBC-RCI	The Link (second hour)	3
2200	CBC-RCI	The Link (second hour)	2-6
0500	RTE-R1	The Weekend on One	1
0500	RTE-R1	The Weekend on One	7
0400	SABC-SAFM	Weekend AM Live	1/7

Health & Medicine Programs

Time	Station/Network	Program Name	Day(s)
0230	ABC-RA	All in the Mind	7
0530	ABC-RA	All in the Mind	7
1130	ABC-RA	All in the Mind	7
1330	ABC-RA	All in the Mind	6
1400	ABC-RA	All in the Mind	4
1700	ABC-RA	All in the Mind	6
0300	ABC-RN	All in the Mind	2/7
1600	ABC-RN	All in the Mind	2
0210	RTI	Health Beats	4
0330	BBCWS-AM	Health Check	3
0830	BBCWS-AM	Health Check	3
1330	BBCWS-AM	Health Check	2
1430	BBCWS-AM	Health Check	2
2230	BBCWS-AM	Health Check	2
1230	BBCWS-IE	Health Check	2
1530	BBCWS-IE	Health Check	2
1930	BBCWS-IE	Health Check	2
2330	BBCWS-IE	Health Check	2
1900	SABC-SAFM	Health Hour	3
0700	WRS-SUI	Health Matters	3
1030	ABC-RA	Health Report	2
1530	ABC-RA	Health Report	2
2230	ABC-RN	Health Report	1
1400	ABC-RN	Health Report (from 2/09)	2
1000	ABC-RN	Health Report (til 1/09)	2
1530	ABC-RN	Health Report (til 1/09)	2
1400	CBC-R1A	White Coat, Black Art	2
1930	CBC-R1A	White Coat, Black Art	7
1600	CBC-R1C	White Coat, Black Art	2
2130	CBC-R1C	White Coat, Black Art	7
1500	CBC-R1E	White Coat, Black Art	2
2030	CBC-R1E	White Coat, Black Art	7
1700	CBC-R1M	White Coat, Black Art	2
2230	CBC-R1M	White Coat, Black Art	7
1800	CBC-R1P	White Coat, Black Art	2
2330	CBC-R1P	White Coat, Black Art	7
1830	CBC-R1S	White Coat, Black Art	7
2030	CBC-R1S	White Coat, Black Art	2

Literature & Drama Programs

Time	Station/Network	Program Name	Day(s)
2145	BBC-R4	A Book at Bedtime	2-6
0200	CBC-R1A	Afghanada	5
1400	CBC-R1A	Afghanada	5
0400	CBC-R1C	Afghanada	5
1600	CBC-R1C	Afghanada	5
0300	CBC-R1E	Afghanada	5
1500	CBC-R1E	Afghanada	5
0500	CBC-R1M	Afghanada	5
1700	CBC-R1M	Afghanada	5
0600	CBC-R1P	Afghanada	5
1800	CBC-R1P	Afghanada	5
0500	ABC-RN	Airplay (from 2/09)	1
0900	ABC-RN	Airplay (from 2/09)	5
1100	ABC-RN	Airplay (from 2/09)	6
0500	ABC-RN	Airplay (til 1/09)	1
1100	ABC-RN	Airplay (til 1/09)	6
1030	BBC-R4	Arts and drama performance series	6
0730	CBC-R1S	Between the Covers	3-7
1730	CBC-R1S	Between the Covers	2-6
1045	RTHK-3	Book Club	1
0845	BBC-R4	Book of the Week	2-6
2330	BBC-R4	Book of the Week	2-6
2245	RTE-R1	Book on One	2-5
0400	ABC-RN	Book Reading	2-6
1300	ABC-RN	Book Reading	2-6
1515	RNZ-NAT	Book reading or Short story	1-7
2230	CRI-WASH	Chinese Writings	1-7
1030	BBC-R4	Comedy series	2
1700	SABC-SAFM	Drama	7
0045	ABC-RN	First Person	2-6
1045	ABC-RN	First Person (from 2/09)	2-5
1445	ABC-RN	First Person (from 2/09)	1
1445	ABC-RN	First Person (til 1/09)	1-5
1640	BBC-R4	From Fact to Fiction	1
1800	BBC-R4	From Fact to Fiction	7
1030	BBC-R4	Inspector Steine	4
1845	BBC-R4	Literature or drama program	1
0345	VOR-WS	Musical Tales	4/7
0845	VOR-WS	Musical Tales	4
0950	VOR-WS	Musical Tales	2
0500	ABC-RN	PoeticA	5/7
1530	BBC-R4	Poetry Please	1
2230	BBC-R4	Poetry Please	7
0000	RNW	Radio Books	7
0030	RNW	Radio Books	2
0030	RNW	Radio Books	5
0100	RNW	Radio Books	7
0130	RNW	Radio Books	5
0200	RNW	Radio Books	7
0230	RNW	Radio Books	2/5
0330	RNW	Radio Books	2/5
0405	RNW	Radio Books	5
0430	RNW	Radio Books	5
0500	RNW	Radio Books	7
0505	RNW	Radio Books	5
0530	RNW	Radio Books	2/5
0600	RNW	Radio Books	7
0605	RNW	Radio Books	5
0630	RNW	Radio Books	5
0700	RNW	Radio Books	7
0705	RNW	Radio Books	5
0730	RNW	Radio Books	2
0730	RNW	Radio Books	5
0800	RNW	Radio Books	7
0805	RNW	Radio Books	5
0830	RNW	Radio Books	5
0900	RNW	Radio Books	4
0930	RNW	Radio Books	6
1000	RNW	Radio Books	4
1030	RNW	Radio Books	1/6
1100	RNW	Radio Books	4
1130	RNW	Radio Books	6
1200	RNW	Radio Books	4
1230	RNW	Radio Books	6
1330	RNW	Radio Books	4

Literature & Drama Programs *continued*

Time	Station/Network	Program Name	Day(s)
1400	RNW	Radio Books	6
1430	RNW	Radio Books	1
1530	RNW	Radio Books	4
1630	RNW	Radio Books	6
1730	RNW	Radio Books	4
1830	RNW	Radio Books	1/6
1930	RNW	Radio Books	4
2030	RNW	Radio Books	1/6
2130	RNW	Radio Books	4
2230	RNW	Radio Books	4
2300	RNW	Radio Books	1/6
2330	RNW	Radio Books	4
1100	SABC-SAFM	SAfm Literature	1
1400	SABC-SAFM	SAfm Sports Special	1
2230	ABC-RN	Short Story	7
1845	RTE-R1	Short Story	1
0535	ABC-RN	Short Story (til 1/09)	1
1050	ABC-RN	Short Story (til 1/09)	4
1800	RNZ-NAT	Storytime	6/7
0810	RTE-R1	Sunday Miscellany	1
1315	BBC-R4	The Afternoon Play	2-6
1430	BBC-R4	The Afternoon Reading	2-6
1300	BBC-R4	The Archers	2-6
1800	BBC-R4	The Archers	1-6
0900	BBC-R4	The Archers Omnibus	1
2245	RNZ-NAT	The Book Reading	1-5
1400	ABC-RN	The Book Show (til 1/09)	1-5
1400	BBC-R4	The Classic Serial	1
2000	BBC-R4	The Classic Serial	7
2330	BBC-R4	The Late Story	7
1035	ABC-RN	The Night Air	1
1135	ABC-RN	The Night Air (til 1/09)	6
1400	ABC-RN	The Night Air (til 1/09)	6
0445	RTE-R1	The Poem and the Place	1
1830	RTE-R1	The Poetry Programme	7
1330	BBC-R4	The Saturday Play	7
0304	RNZ-NAT	The Sunday Drama	1
0904	RNZ-NAT	The Wednesday Drama	4
1030	BBC-R4	With Great Pleasure	5
1845	BBC-R4	Woman's Hour Drama	2-6
1900	WRS-SUI	World Drama	7

Light Entertainment, Humor, Quiz, & Panel Game Programs

Time	Station/Network	Program Name	Day(s)
1930	ABC-RN	BBC Comedy	1/2/6
0204	RNZ-NAT	Comedy	1
1730	BBC-R4	Comedy series	3-5
2200	BBC-R4	Comedy series	3-5
0800	BBC-R4	Desert Island Discs	6
1015	BBC-R4	Desert Island Discs	1
1800	CBC-R1A	Festival of Funny	6
2000	CBC-R1C	Festival of Funny	6
1900	CBC-R1E	Festival of Funny	6
2100	CBC-R1M	Festival of Funny	6
2200	CBC-R1P	Festival of Funny	6
0900	ABC-RN	Garrison Keillor's Radio Show	1
0030	RTHK-3	Give Me Five	1
1300	CBC-R1A	Go!	7
1500	CBC-R1C	Go!	7
1400	CBC-R1E	Go!	7
1600	CBC-R1M	Go!	7
1700	CBC-R1P	Go!	7
1300	CBC-R1S	Go!	7
2300	CBC-R1S	Go!	7
0235	RTI	Instant Noodles	5
1800	CBC-R1A	Laugh Out Loud	2
2230	CBC-R1A	Laugh Out Loud	7
2000	CBC-R1C	Laugh Out Loud	2
2330	CBC-R1C	Laugh Out Loud	7

Light Entertainment, Humor, Quiz, & Panel Game Programs *continued*

Time	Station/Network	Program Name	Day(s)
1900	CBC-R1E	Laugh Out Loud	2
2230	CBC-R1E	Laugh Out Loud	7
0030	CBC-R1M	Laugh Out Loud	1
2100	CBC-R1M	Laugh Out Loud	2
0130	CBC-R1P	Laugh Out Loud	1
2200	CBC-R1P	Laugh Out Loud	2
1430	CBC-R1S	Laugh Out Loud	7
2030	CBC-R1S	Laugh Out Loud	5
1230	RNZ-NAT	Laugh Track	6
1715	BBC-R4	Loose Ends	7
1930	ABC-RN	My Music	3
1930	ABC-RN	My Word	4
1100	BBC-R4	Quiz or panel game	1
1230	BBC-R4	Quiz or panel game	2
1730	BBC-R4	Quiz or panel game	2
2200	BBC-R4	Quiz or panel game	7
1930	ABC-RN	Radio National Quiz	7
0700	RNZ-NAT	Saturday Night with Peter Fry	7
1430	CBC-R1A	The Debaters	7
2330	CBC-R1A	The Debaters	1
0130	CBC-R1C	The Debaters	2
1630	CBC-R1C	The Debaters	7
0030	CBC-R1E	The Debaters	2
1530	CBC-R1E	The Debaters	7
0230	CBC-R1M	The Debaters	2
1730	CBC-R1M	The Debaters	7
0330	CBC-R1P	The Debaters	2
1830	CBC-R1P	The Debaters	7
1930	ABC-RN	The Goons	5
1130	BBC-R4	The News Quiz	7
1730	BBC-R4	The News Quiz	6
0200	CBC-R1A	The Vinyl Cafe	4
1500	CBC-R1A	The Vinyl Cafe	1
0400	CBC-R1C	The Vinyl Cafe	4
1700	CBC-R1C	The Vinyl Cafe	1
0300	CBC-R1E	The Vinyl Cafe	4
1600	CBC-R1E	The Vinyl Cafe	1
0500	CBC-R1M	The Vinyl Cafe	4
1800	CBC-R1M	The Vinyl Cafe	1
0600	CBC-R1P	The Vinyl Cafe	4
1900	CBC-R1P	The Vinyl Cafe	1
0000	CBC-R1S	The Vinyl Cafe	7
0800	CBC-R1S	The Vinyl Cafe	7
1000	CBC-R1S	The Vinyl Cafe	6
1900	CBC-R1S	The Vinyl Cafe	1
0230	CBC-R1A	WireTap	5
1600	CBC-R1A	WireTap	1
0430	CBC-R1C	WireTap	5
1800	CBC-R1C	WireTap	1
0330	CBC-R1E	WireTap	5
1700	CBC-R1E	WireTap	1
0530	CBC-R1M	WireTap	5
1900	CBC-R1M	WireTap	1
0630	CBC-R1P	WireTap	5
2300	CBC-R1P	WireTap	1
0030	CBC-R1S	WireTap	1
0130	CBC-R1S	WireTap	2
2030	CBC-R1S	WireTap	6

Listener Interaction & Mailbag Programs

Time	Station/Network	Program Name	Day(s)
1430	RFI	Club 9516	1
1230	BBC-R4	Feedback	6
1900	BBC-R4	Feedback	1
0415	PR-EXT	In Touch	3
0750	PR-EXT	In Touch	3
1230	PR-EXT	In Touch	1
1730	PR-EXT	In Touch	6
2000	PR-EXT	In Touch	6

Listener Interaction & Mailbag Programs *continued*

Time	Station/Network	Program Name	Day(s)
0305	DW	In-Box	1
0405	DW	In-Box	1
0605	DW	In-Box	1
0705	DW	In-Box	1
0905	DW	In-Box	1
1105	DW	In-Box	1
1305	DW	In-Box	1
1505	DW	In-Box	1
0027	CRI-ENG	Listeners' Garden	7
0127	CRI-ENG	Listeners' Garden	7
0327	CRI-ENG	Listeners' Garden	7
0427	CRI-ENG	Listeners' Garden	7
0527	CRI-ENG	Listeners' Garden	7
0627	CRI-ENG	Listeners' Garden	7
1327	CRI-ENG	Listeners' Garden	6
1427	CRI-ENG	Listeners' Garden	6
1527	CRI-ENG	Listeners' Garden	6
1330	CRI-WASH	Listeners' Garden	2
0030	CRI-RTC	Listeners' Garden	7
0130	CRI-RTC	Listeners' Garden	7
0230	CRI-RTC	Listeners' Garden	7
0330	CRI-RTC	Listeners' Garden	7
0430	CRI-RTC	Listeners' Garden	7
0530	CRI-RTC	Listeners' Garden	7
0630	CRI-RTC	Listeners' Garden	7
0830	CRI-RTC	Listeners' Garden	7
0930	CRI-RTC	Listeners' Garden	7
1230	CRI-RTC	Listeners' Garden	7
1330	CRI-RTC	Listeners' Garden	7
1430	CRI-RTC	Listeners' Garden	7
1530	CRI-RTC	Listeners' Garden	7
1630	CRI-RTC	Listeners' Garden	7
1830	CRI-RTC	Listeners' Garden	7
1930	CRI-RTC	Listeners' Garden	7
2030	CRI-RTC	Listeners' Garden	7
2130	CRI-RTC	Listeners' Garden	7
2330	CRI-RTC	Listeners' Garden	6
0230	RTE-R1	Liveline	3-7
1245	RTE-R1	Liveline	2-6
0105	R.PRG	Mailbox	2
0205	R.PRG	Mailbox	2
0405	R.PRG	Mailbox	2
0435	R.PRG	Mailbox	2
0805	R.PRG	Mailbox	1
1135	R.PRG	Mailbox	1
1405	R.PRG	Mailbox	1
1705	R.PRG	Mailbox	1
1805	R.PRG	Mailbox	1
2105	R.PRG	Mailbox	1
2235	R.PRG	Mailbox	1
2335	R.PRG	Mailbox	1
0200	VOR-WS	Moscow Mailbag	1/2
0400	VOR-WS	Moscow Mailbag	4/7
0500	VOR-WS	Moscow Mailbag	3/6
0600	VOR-WS	Moscow Mailbag	2
0700	VOR-WS	Moscow Mailbag	1/5
0800	VOR-WS	Moscow Mailbag	6
0140	BBCWS-AM	Over to You	2
0640	BBCWS-AM	Over to You	1
2140	BBCWS-AM	Over to You	1
0240	BBCWS-IE	Over to You	2
1040	BBCWS-IE	Over to You	1
1840	BBCWS-IE	Over to You	7
2240	BBCWS-IE	Over to You	7
1300	WRN-NA	RTE Ireland: Sunday Miscellany	1
1000	CBC-RCI	The Maple Leaf Mailbag	2
1500	CBC-RCI	The Maple Leaf Mailbag	1
1505	CBC-RCI	The Maple Leaf Mailbag	1
2200	CBC-RCI	The Maple Leaf Mailbag	1
2300	CBC-RCI	The Maple Leaf Mailbag	1
2305	CBC-RCI	The Maple Leaf Mailbag	1
0210	RTI	We've Got Mail!	3
0010	NHK-RJ	World Interactive	1

Listener Interaction & Mailbag Programs continued

Time	Station/Network	Program Name	Day(s)
0510	NHK-RJ	World Interactive	7
1210	NHK-RJ	World Interactive	7
1410	NHK-RJ	World Interactive	7
1700	BBCWS-AM	World, Have Your Say	2-6
1700	BBCWS-IE	World, Have Your Say	2-6
1700	BBCWS-NX	World, Have Your Say	2-6
1700	BBCWS-PR	World, Have Your Say	2-6
0240	KBS-WR	Worldwide Friendship	7
0315	KBS-WR	Worldwide Friendship	7
0510	KBS-WR	Worldwide Friendship	7
0640	KBS-WR	Worldwide Friendship	7
0810	KBS-WR	Worldwide Friendship	7
0840	KBS-WR	Worldwide Friendship	7
1110	KBS-WR	Worldwide Friendship	7
1210	KBS-WR	Worldwide Friendship	7
1210	KBS-WR	Worldwide Friendship	7
1510	KBS-WR	Worldwide Friendship	7
2040	KBS-WR	Worldwide Friendship	7
2215	KBS-WR	Worldwide Friendship	7
2315	KBS-WR	Worldwide Friendship	7

Music Programs

Time	Station/Network	Program Name	Day(s)
2100	CBC-R1A	A Propos	7
0200	CBC-R1C	A Propos	1
0100	CBC-R1E	A Propos	1
0300	CBC-R1M	A Propos	1
0400	CBC-R1P	A Propos	1
0530	DW	A World of Music	7
0730	DW	A World of Music	1
0830	DW	A World of Music	7
1030	DW	A World of Music	7
1130	DW	A World of Music	1
1330	DW	A World of Music	1
1530	DW	A World of Music	1
1730	DW	A World of Music	1
2230	DW	A World of Music	1
0230	DW	A World of Music	2
1100	SABC-SAFM	African Connection	7
0315	SABC-CHAF	African Music	2
0440	SABC-CHAF	African Music	3
0640	SABC-CHAF	African Music	2
1040	SABC-CHAF	African Music	2
1415	SABC-CHAF	African Music	2
1515	SABC-CHAF	African Music	2
1015	SABC-CHAF	African Music	7
0000	SABC-CHAF	African Music (News on the hour)	1-7
2100	SABC-CHAF	African Music (News on the hour)	1
1400	RTHK-3	All the Way with Ray	2-6
0515	RTHK-3	Alyson Hau	7
0230	ABC-RA	Australian Country Style	1
0530	ABC-RA	Australian Country Style	1
2030	ABC-RA	Australian Country Style	6
2212	RTE-R1	Balfe's Sunday Best	1
1100	RNZ-NAT	Beale Street Caravan	2
0443	BBC-R4	Bells on Sunday	1
2345	BBC-R4	Bells on Sunday	1
1900	SABC-SAFM	Best of Jazz	7
0800	CRI-WASH	Beyond Beijing	1-7
1700	CRI-WASH	Beyond Beijing	1-7
2000	RTE-R1	Ceili House	7
0330	BBCWS-AM	Charlie Gillett's World of Music	2
2230	BBCWS-AM	Charlie Gillett's World of Music	7
2330	BBCWS-AM	Charlie Gillett's World of Music	1
0130	BBCWS-IE	Charlie Gillett's World of Music	1
2130	BBCWS-IE	Charlie Gillett's World of Music	1
0130	BBCWS-PR	Charlie Gillett's World of Music	1
0630	BBCWS-PR	Charlie Gillett's World of Music	1
1730	BBCWS-PR	Charlie Gillett's World of Music	7
1100	RNZ-NAT	Charlie Gillett's World of Music	3

Music Programs continued

Time	Station/Network	Program Name	Day(s)
0405	PR-EXT	Chart Show	6
0740	PR-EXT	Chart Show	6
1240	PR-EXT	Chart Show	7
1735	PR-EXT	Chart Show	1
2005	PR-EXT	Chart Show	1
1127	CRI-ENG	China Beat	1
2327	CRI-ENG	China Beat	7
0727	CRI-RTC	China Beat	1/7
1027	CRI-RTC	China Beat	1/7
1127	CRI-RTC	China Beat	1/7
1727	CRI-RTC	China Beat	1/7
2227	CRI-RTC	China Beat	1/7
0200	CRI-WASH	China Beat	2/3
1400	CRI-WASH	China Beat	1/2
1550	CRI-WASH	Chinese Melody	1-7
0715	SABC-CHAF	Choral Music	1
1000	SABC-SAFM	Choral Music	1
1900	SABC-SAFM	Classical Sunday	1
0805	DW	Concert Hour	1
1005	DW	Concert Hour	1
1205	DW	Concert Hour	1
1405	DW	Concert Hour	1
1400	RTHK-3	Cool Trax	7
2200	RTE-R1	Country Time	7
0200	CBC-R1S	Deep Roots	2
0700	CBC-R1S	Deep Roots	2
2200	ORF-FM4	Digital Konfusion Mixshow (in German/English)	7
0200	SABC-SAFM	Early Classics	1/7
2200	RTHK-3	Early Show	6
2100	RTE-R1	Failte Isteach	7
0330	VOR-WS	Folk Box	3
0730	VOR-WS	Folk Box	7
0830	VOR-WS	Folk Box	3
0930	VOR-WS	Folk Box	5
0220	RTI	Groove Zone	7
1430	RNZ-NAT	Hidden Treasures with Trevor Reekie	1
2230	RNZ-NAT	Hidden Treasures with Trevor Reekie	7
0405	PR-EXT	High Note	2
0730	PR-EXT	High Note	2
1735	PR-EXT	High Note	5
2005	PR-EXT	High Note	5
0530	DW	Hits in Germany	3
0730	DW	Hits in Germany	3
0930	DW	Hits in Germany	3
1130	DW	Hits in Germany	3
1330	DW	Hits in Germany	3
1530	DW	Hits in Germany	3
1730	DW	Hits in Germany	3
1930	DW	Hits in Germany	3
2130	DW	Hits in Germany	3
2330	DW	Hits in Germany	3
0100	CBC-R1A	In the Key of Charles	2
0200	CBC-R1C	In the Key of Charles	2
0100	CBC-R1E	In the Key of Charles	2
0300	CBC-R1M	In the Key of Charles	2
0400	CBC-R1P	In the Key of Charles	2
0000	CBC-R1A	Inside the Music	2
0100	CBC-R1C	Inside the Music	2
0000	CBC-R1E	Inside the Music	2
0200	CBC-R1M	Inside the Music	2
0300	CBC-R1P	Inside the Music	2
0800	CBC-R1S	Inside the Music	2
1000	CBC-R1S	Inside the Music	1
0500	ABC-RN	Into the Music	6
0700	ABC-RN	Into the Music	7
0235	RTI	Jade Bells and Bamboo Pipes	4
0930	ABC-RA	Jazz Notes	7
0330	VOR-WS	Jazz Show	6
0730	VOR-WS	Jazz Show	4/5
0930	VOR-WS	Jazz Show	6
0240	KBS-WR	Korean Pop Interactive	1
0315	KBS-WR	Korean Pop Interactive	1
0510	KBS-WR	Korean Pop Interactive	1
0640	KBS-WR	Korean Pop Interactive	1
0810	KBS-WR	Korean Pop Interactive	1

Time	Station/Network	Program Name	Day(s)
0840	KBS-WR	Korean Pop Interactive	1
1110	KBS-WR	Korean Pop Interactive	1
1210	KBS-WR	Korean Pop Interactive	1
1210	KBS-WR	Korean Pop Interactive	1
1510	KBS-WR	Korean Pop Interactive	1
2040	KBS-WR	Korean Pop Interactive	1
2215	KBS-WR	Korean Pop Interactive	1
2315	KBS-WR	Korean Pop Interactive	1
2200	RTE-R1	Late Date	6
2300	RTE-R1	Late Date	7-5
1600	SABC-SAFM	Living Sounds	1
0119	R.PRG	Magic Carpet (monthly)	2
0219	R.PRG	Magic Carpet (monthly)	2
0419	R.PRG	Magic Carpet (monthly)	2
0449	R.PRG	Magic Carpet (monthly)	2
0819	R.PRG	Magic Carpet (monthly)	1
1149	R.PRG	Magic Carpet (monthly)	1
1419	R.PRG	Magic Carpet (monthly)	1
1719	R.PRG	Magic Carpet (monthly)	1
1819	R.PRG	Magic Carpet (monthly)	1
2119	R.PRG	Magic Carpet (monthly)	1
2249	R.PRG	Magic Carpet (monthly)	1
2349	R.PRG	Magic Carpet (monthly)	1
1015	SABC-CHAF	Modern Africa Music	1
0400	ORF-FM4	Morning Show (in German and English)	1-7
1930	WRS-SUI	Music	2-6
2100	WRS-SUI	Music	1-7
2255	CRI-RTC	Music	1-7
0200	RNZ-NAT	Music 101 with Kirsten Johnstone	7
0400	VOR-WS	Music and Musicians	1
0500	VOR-WS	Music and Musicians	5
0800	VOR-WS	Music and Musicians	2
0550	VOR-WS	Music Around Us	3
0650	VOR-WS	Music Around Us	6
0530	VOR-WS	Music Calendar	3
0630	VOR-WS	Music Calendar	6
0600	ABC-RN	Music Deli	1
1800	ABC-RN	Music Deli	6
1000	ABC-RN	Music Deli	6
1345	BBC-R4	Music documentary or feature series	1
0400	RNZ-NAT	Music feature	7
0730	RNZ-NAT	Music feature	6
1204	RNZ-NAT	Music from Midnight	1-7
1127	CRI-ENG	Music Memories	7
2327	CRI-ENG	Music Memories	6
0119	R.PRG	Music Profile (monthly)	2
0219	R.PRG	Music Profile (monthly)	2
0419	R.PRG	Music Profile (monthly)	2
0449	R.PRG	Music Profile (monthly)	2
0819	R.PRG	Music Profile (monthly)	1
1149	R.PRG	Music Profile (monthly)	1
1419	R.PRG	Music Profile (monthly)	1
1719	R.PRG	Music Profile (monthly)	1
1819	R.PRG	Music Profile (monthly)	1
2119	R.PRG	Music Profile (monthly)	1
2249	R.PRG	Music Profile (monthly)	1
2349	R.PRG	Music Profile (monthly)	1
0330	RNZI	New Music Releases	2
0730	RNZI	New Music Releases	2
1100	RNZ-NAT	New Zealand Music Festival or Feature	6
1800	WRS-SUI	News and Music	7
1900	WRS-SUI	News and Music	1
1800	RTHK-3	Night Music	1-5
1800	RTHK-3	Night Music	6/7
2000	RTE-R1	O'Brien on Song	1
0800	RNZI	Pacific Music	2-6
1200	RTHK-3	Pete's Quiet Night In	1
1100	RTHK-3	Peter King	2-6
1405	RNZ-NAT	Playing Favourites	4
0010	NHK-RJ	Pop Up Japan	2
0510	NHK-RJ	Pop Up Japan	1
1210	NHK-RJ	Pop Up Japan	1
1410	NHK-RJ	Pop Up Japan	1
1600	ABC-RN	Quiet Space	7
0430	RTE-R1	Risin' Time	2-6

Music Programs *continued*

Time	Station/Network	Program Name	Day(s)
1100	RNZ-NAT	Round Midnight with Martin Kwok	4
0450	VOR-WS	Russia--1000 Years of Music	6
0350	VOR-WS	Russian Hits	1
0850	VOR-WS	Russian Hits	1
2200	SABC-SAFM	SAfm Twilights	1-7
0100	RTHK-3	Saturday Morning with Phil	7
0100	CBC-R1A	Saturday Night Blues	1
0300	CBC-R1C	Saturday Night Blues	1
0200	CBC-R1E	Saturday Night Blues	1
0400	CBC-R1M	Saturday Night Blues	1
0500	CBC-R1P	Saturday Night Blues	1
1200	ABC-RA	Saturday Night Country	7
0800	RTHK-3	Saturday PM	7
0515	RTHK-3	Simon Willson	1
2300	ORF-FM4	Sleepless (in German and English)	2-5
0345	VOR-WS	Songs from Russia	1
0845	VOR-WS	Songs from Russia	1
1230	BBC-R4	Soul Music	3
1430	BBC-R4	Soul Music	7
1320	ABC-RN	Sound Quality	6
1700	ABC-RN	Sound Quality	1
1400	ABC-RN	Sound Quality (from 2/09)	6
0230	RTI	Soundwaves	2
1900	RTE-R1	South Wind Blows	7
0700	RTHK-3	Steve James	2-6
2200	RTHK-3	Sunday Early Show	7
1400	RTHK-3	Sunday Late	1
0800	ORF-FM4	Sunny Side Up (in German and English)	1
0340	PR-EXT	Talking Jazz	3
0715	PR-EXT	Talking Jazz	3
1230	PR-EXT	Talking Jazz	2/5
1730	PR-EXT	Talking Jazz	2
2000	PR-EXT	Talking Jazz	2
0420	ABC-RN	The Daily Planet	2-6
1320	ABC-RN	The Daily Planet	2-5
0300	RTE-R1	The Late Session	1
2100	RTE-R1	The Late Session	1
1100	RNZ-NAT	The Music Mix	5
0900	ABC-RA	The Music Show	1
0000	ABC-RN	The Music Show	7
1000	ABC-RN	The Music Show	7
0300	RTE-R1	The Radio 1 Music Collection	3-7
2000	RTE-R1	The Radio 1 Music Collection	2-6
0935	ABC-RN	The Rhythm Divine (from 2/09)	6
1100	RTE-R1	The Ronan Collins Show	2-6
0730	RNZ-NAT	The Sampler	3
1430	RNZ-NAT	The Sampler	5
1200	ABC-RN	The Weekend Planet	1/7
0100	BBCWS-AM	The World Today	1
0300	CBC-R1A	Tonic	2
0400	CBC-R1C	Tonic	2
0300	CBC-R1E	Tonic	2
0500	CBC-R1M	Tonic	2
0600	CBC-R1P	Tonic	2
0800	ORF-FM4	Update (in German and English)	2-7
1415	SABC-CHAF	Variety Music	7
0200	CBC-R1A	Vinyl Tap	7
2300	CBC-R1A	Vinyl Tap	7
0000	CBC-R1C	Vinyl Tap	1
0400	CBC-R1C	Vinyl Tap	7
0300	CBC-R1E	Vinyl Tap	7
2300	CBC-R1E	Vinyl Tap	7
0100	CBC-R1M	Vinyl Tap	1
0500	CBC-R1M	Vinyl Tap	7
0200	CBC-R1P	Vinyl Tap	1
0600	CBC-R1P	Vinyl Tap	7
0300	CBC-R1S	Vinyl Tap	2
0700	CBC-R1S	Vinyl Tap	1
2000	CBC-R1S	Vinyl Tap	7
1430	RNZ-NAT	Waiata	6
1430	RNZ-NAT	Waiata	6
1130	RNZI	Waiata	6
1330	RNZI	Waiata	6
1630	RNZI	Waiata	4/6
1040	RNZ-NAT	Wayne's Music	1

Music Programs continued

Time	Station/Network	Program Name	Day(s)
1100	RNZ-NAT	Wayne's Music	7
0730	RFI	World Tracks	6
1430	RFI	World Tracks	6
1200	RTHK-3	World Vibes	7
1500	ORF-FM4	World Wide Show (in German and English)	1

News Programs

Time	Station/Network	Program Name	Day(s)
0420	RFI	Africa Report	2-6
0520	RFI	Africa Report	2-6
0720	RFI	Africa Report	2-6
0615	SABC-CHAF	Africa Rise and Shine	3-6
1715	SABC-CHAF	Africa This Week	7
1200	SABC-SAFM	Afternoon Talk	2-6
0140	BBCWS-AM	Analysis	3-6
0440	BBCWS-AM	Analysis	2-6
0749	BBCWS-AM	Analysis	2-6
1040	BBCWS-AM	Analysis	2-6
2140	BBCWS-AM	Analysis	2-6
2240	BBCWS-AM	Analysis	1
0040	BBCWS-IE	Analysis	2-6
0440	BBCWS-IE	Analysis	2-6
0749	BBCWS-IE	Analysis	2-6
1040	BBCWS-IE	Analysis	2-6
2120	BBCWS-IE	Analysis	2-6
0040	BBCWS-NX	Analysis	2-6
0140	BBCWS-NX	Analysis	2-6
0440	BBCWS-NX	Analysis	2-6
0750	BBCWS-NX	Analysis	2-6
0840	BBCWS-NX	Analysis	2-6
1040	BBCWS-NX	Analysis	2-6
1140	BBCWS-NX	Analysis	2-6
1440	BBCWS-NX	Analysis	2-6
1840	BBCWS-NX	Analysis	2-6
1940	BBCWS-NX	Analysis	2-6
2140	BBCWS-NX	Analysis	2-6
2240	BBCWS-NX	Analysis	1-5
0040	BBCWS-PR	Analysis	2-6
1450	BBCWS-PR	Analysis	2-6
1540	BBCWS-PR	Analysis	2-6
2240	BBCWS-PR	Analysis	1-5
0130	CBC-R1S	As It Happens	3
0130	CBC-R1S	As It Happens	4-7
2130	CBC-R1S	As It Happens	2
2130	CBC-R1S	As It Happens	3-6
1415	RFI	Asia Pacific	7
0000	BBCWS-AM	Assignment	6
0800	BBC-R4	Broadcasting House	1
0705	ABC-RA	Correspondent's Notebook	1
2100	ABC-RA	Correspondent's Notebook	6
0800	ABC-RA	Correspondents Report	1
1800	ABC-RA	Correspondents Report	7
2200	ABC-RA	Correspondents Report	7
2200	ABC-RN	Correspondents Report	7
1045	WRS-SUI	Cover Story	2-6
0715	SABC-CHAF	Current Affairs	3
1015	SABC-CHAF	Current Affairs	3-6
1415	SABC-CHAF	Current Affairs	3-6
1515	SABC-CHAF	Current Affairs	3-6
1715	SABC-CHAF	Current Affairs	2-6
0138	RNZI	Dateline Pacific	2-6
0308	RNZI	Dateline Pacific	2-7
0708	RNZI	Dateline Pacific	2-6
0838	RNZI	Dateline Pacific	2-6
1108	RNZI	Dateline Pacific	2-6
1308	RNZI	Dateline Pacific	2-6
1608	RNZI	Dateline Pacific	2-6
1815	RNZI	Dateline Pacific	2-6
2015	RNZI	Dateline Pacific	1-6
2215	RNZI	Dateline Pacific	1-5
2230	CBC-R1A	Dispatches	1
2230	CBC-R1A	Dispatches	2

News Programs continued

Time	Station/Network	Program Name	Day(s)
0030	CBC-R1C	Dispatches	3
2330	CBC-R1C	Dispatches	1
2230	CBC-R1E	Dispatches	1
2330	CBC-R1E	Dispatches	2
0030	CBC-R1M	Dispatches	2
0130	CBC-R1M	Dispatches	3
0130	CBC-R1P	Dispatches	2
0230	CBC-R1P	Dispatches	3
0000	CBC-R1S	Dispatches	5
0800	CBC-R1S	Dispatches	5
1000	CBC-R1S	Dispatches	4
2300	CBC-R1S	Dispatches	1
1700	SABC-SAFM	Evening Talk	2-6
0600	VOR-WS	Focus on Asia and the Pacific	3-7
1000	BBC-R4	From Our Own Correspondent	5
1030	BBC-R4	From Our Own Correspondent	7
0100	BBCWS-AM	From Our Own Correspondent	7
0430	BBCWS-AM	From Our Own Correspondent	7
0830	BBCWS-AM	From Our Own Correspondent	2
1300	BBCWS-AM	From Our Own Correspondent	1
1800	BBCWS-AM	From Our Own Correspondent	7
2200	BBCWS-AM	From Our Own Correspondent	7
0100	BBCWS-IE	From Our Own Correspondent	7
0500	BBCWS-IE	From Our Own Correspondent	1
1030	BBCWS-IE	From Our Own Correspondent	7
2200	BBCWS-IE	From Our Own Correspondent	7
0100	BBCWS-NX	From Our Own Correspondent	7
0430	BBCWS-NX	From Our Own Correspondent	7
0800	BBCWS-NX	From Our Own Correspondent	7
1030	BBCWS-NX	From Our Own Correspondent	1
1400	BBCWS-NX	From Our Own Correspondent	1
1800	BBCWS-NX	From Our Own Correspondent	7
2200	BBCWS-NX	From Our Own Correspondent	7
0100	BBCWS-PR	From Our Own Correspondent	7
0430	BBCWS-PR	From Our Own Correspondent	1
0730	BBCWS-PR	From Our Own Correspondent	2
0800	BBCWS-PR	From Our Own Correspondent	7
1130	BBCWS-PR	From Our Own Correspondent	1
1400	BBCWS-PR	From Our Own Correspondent	1
1800	BBCWS-PR	From Our Own Correspondent	7
1830	BBCWS-PR	From Our Own Correspondent	2
2200	BBCWS-PR	From Our Own Correspondent	7
0300	CBC-R1A	From Our Own Correspondent	3
0500	CBC-R1C	From Our Own Correspondent	3
0400	CBC-R1E	From Our Own Correspondent	3
0600	CBC-R1M	From Our Own Correspondent	3
0700	CBC-R1P	From Our Own Correspondent	3
0230	DW	Insight	1
0430	DW	Insight	7
0530	DW	Insight	1
0630	DW	Insight	7
0930	DW	Insight	7
1230	DW	Insight	7
1430	DW	Insight	7
1630	DW	Insight	1
1730	DW	Insight	7
1830	DW	Insight	1
2230	DW	Insight	7
0904	RNZ-NAT	Insight	2
1230	RNZ-NAT	Insight	3
2012	RNZ-NAT	Insight	7
0700	SABC-SAFM	Morning Talk	2-6
1600	ABC-RA	National Interest	6
0705	PR-EXT	Network Europe Week	1
1205	PR-EXT	Network Europe Week	7
0300	RNW	Network Europe Week	1
1305	RNW	Network Europe Week	7
1500	RNW	Network Europe Week	7
1805	RNW	Network Europe Week	7
0300	VOR-WS	News and Views	3-1
0900	VOR-WS	News and Views	3-7
0210	RTI	News Talk	6
0330	RNZI	Pacific Correspondent	6
0730	RNZI	Pacific Correspondent	6

News Programs *continued*

Time	Station/Network	Program Name	Day(s)
1130	RNZI	Pacific Correspondent	5
1330	RNZI	Pacific Correspondent	5
1630	RNZI	Pacific Correspondent	5
1708	RNZI	Pacific Correspondent	5
2115	RNZI	Pacific Correspondent	5
0135	ABC-RA	Rear Vision	7
0730	ABC-RA	Rear Vision	1
1330	ABC-RA	Rear Vision	5
1700	ABC-RA	Rear Vision	5
1734	ABC-RA	Rear Vision	1
1900	ABC-RN	Rear Vision	1
1030	ABC-RA	Rear Vision (from 2/09)	4
0330	ABC-RN	Rear Vision (from 2/09)	1
1400	ABC-RN	Rear Vision (from 2/09)	4
2230	ABC-RN	Rear Vision (from 2/09)	3
0300	ABC-RN	Rear Vision (til 1/09)	1
1035	ABC-RN	Rear Vision (til 1/09)	3
1530	ABC-RA	Rear Vision (til 2/09)	4
1030	RTHK-3	Reflections from Asia	1
2330	RTHK-3	Reflections from Asia	6
0200	VOR-WS	Russia and the World	3-7
0800	VOR-WS	Russia and the World	3/5/7
0315	SABC-CHAF	Straight Talk	1
0415	SABC-CHAF	Straight Talk	1
0300	VOR-WS	Sunday Panorama	2
1100	ABC-RA	Sunday Profile	1
0300	ABC-RA	Sunday Profile (from 2/09)	1
0600	ABC-RA	Sunday Profile (from 2/09)	1
0300	ABC-RN	Sunday Profile (from 2/09)	1
1130	CBC-R1A	The Current	2-6
1330	CBC-R1C	The Current	2-6
1230	CBC-R1E	The Current	2-6
1430	CBC-R1M	The Current	2-6
1530	CBC-R1P	The Current	2-6
1137	CBC-R1S	The Current	2-6
1437	CBC-R1S	The Current	2-6
0130	BBCWS-AM	The Instant Guide	2
0630	BBCWS-AM	The Instant Guide	1
2130	BBCWS-AM	The Instant Guide	1
0230	BBCWS-IE	The Instant Guide	2
1030	BBCWS-IE	The Instant Guide	1
1830	BBCWS-IE	The Instant Guide	7
2230	BBCWS-IE	The Instant Guide	7
0020	BBCWS-PR	The Instant Guide	1
0420	BBCWS-PR	The Instant Guide	1
1020	BBCWS-PR	The Instant Guide	1
2206	RNZ-NAT	The Sunday Group	7
1535	RNZ-NAT	The Week	6
0200	RTE-R1	This Week	2
1200	RTE-R1	This Week	1
1935	RNZ-NAT	Weekend Worldwatch	7
1935	RNZI	World Watch	7

News Documentary Programs

Time	Station/Network	Program Name	Day(s)
0500	BBCWS-AM	Assignment	6
0800	BBCWS-AM	Assignment	1
1400	BBCWS-AM	Assignment	5
1900	BBCWS-AM	Assignment	5
2300	BBCWS-AM	Assignment	7
0500	BBCWS-IE	Assignment	7
0900	BBCWS-IE	Assignment	5
1200	BBCWS-IE	Assignment	5
1900	BBCWS-IE	Assignment	5
2300	BBCWS-IE	Assignment	5
0100	BBCWS-NX	Assignment	1
0800	BBCWS-NX	Assignment	1
1300	BBCWS-NX	Assignment	7
2100	BBCWS-NX	Assignment	7
0100	BBCWS-PR	Assignment	1
0800	BBCWS-PR	Assignment	1

News Documentary Programs *continued*

Time	Station/Network	Program Name	Day(s)
1300	BBCWS-PR	Assignment	7
1600	BBCWS-PR	Assignment	7
2100	BBCWS-PR	Assignment	7
2300	ABC-RA	Background Briefing	7
0900	ABC-RN	Background Briefing	3
1800	ABC-RN	Background Briefing	3
2300	ABC-RN	Background Briefing	7
1000	BBC-R4	Documentary feature or series	6
1230	BBC-R4	Documentary feature or series	1
1445	BBC-R4	Documentary feature or series	2-6
1900	BBC-R4	Documentary feature or series	2/5
1945	BBC-R4	Documentary feature or series	4
2000	BBC-R4	Documentary feature or series	6
0000	BBCWS-AM	Documentary feature or series	3/5/7
0300	BBCWS-AM	Documentary feature or series	1
0500	BBCWS-AM	Documentary feature or series	3/5/7
0800	BBCWS-AM	Documentary feature or series	7
1400	BBCWS-AM	Documentary feature or series	2/4/6
1500	BBCWS-AM	Documentary feature or series	1
1900	BBCWS-AM	Documentary feature or series	2/4/6-7
2100	BBCWS-AM	Documentary feature or series	1/7
2300	BBCWS-AM	Documentary feature or series	1
0900	BBCWS-IE	Documentary feature or series	2/4/6
1000	BBCWS-IE	Documentary feature or series	1/7
1200	BBCWS-IE	Documentary feature or series	2/4/6
1700	BBCWS-IE	Documentary feature or series	1
1800	BBCWS-IE	Documentary feature or series	7
1900	BBCWS-IE	Documentary feature or series	2/4/6
2100	BBCWS-IE	Documentary feature or series	7
2300	BBCWS-IE	Documentary feature or series	2/4/6
1800	RTE-R1	Documentary on One	1
0300	ABC-RN	Feature program (from 2/09)	4
0400	ABC-RN	Feature program (from 2/09)	7
1600	BBC-R4	File on 4	1
1900	BBC-R4	File on 4	3
0800	ABC-RN	FORA Radio (from 2/09)	2
1600	ABC-RN	Radio Eye	4
0300	ABC-RN	Radio Eye (til 1/09)	4
0400	ABC-RN	Radio Eye (til 1/09)	7
1800	WRN-NA	RTE Ireland: Documentaries	1/7
0815	RNZ-NAT	Windows on the World	2-5

News Magazine Programs

Time	Station/Network	Program Name	Day(s)
0715	SABC-CHAF	37 Degrees	2
0415	SABC-CHAF	Africa Rise and Shine	2/3
0415	SABC-CHAF	Africa Rise and Shine	4-6
0515	SABC-CHAF	Africa Rise and Shine	3-6
1515	SABC-CHAF	Africa This Week	1/7
2100	ABC-RA	AM	1-5
2200	ABC-RA	AM	1-5
2100	ABC-RN	AM	1-5
2130	CBC-R1A	As It Happens	2
2130	CBC-R1A	As It Happens	3-6
2330	CBC-R1C	As It Happens	2
2330	CBC-R1C	As It Happens	3-6
2230	CBC-R1E	As It Happens	2
2230	CBC-R1E	As It Happens	3-6
0030	CBC-R1M	As It Happens	3
0030	CBC-R1M	As It Happens	4-7
0130	CBC-R1P	As It Happens	3
0130	CBC-R1P	As It Happens	4-7
1000	ABC-RA	Asia Pacific	2-6
1300	ABC-RA	Asia Pacific	2-6
1500	ABC-RA	Asia Pacific	2-6
1900	ABC-RN	Asia Pacific	2-6
1430	ABC-RN	Asia Pacific (from 2/09)	2-5
1500	ABC-RN	Asia Pacific (til 1/09)	2-6
2000	WRS-SUI	BBC Newshour	1-7
2200	WRS-SUI	BBC Newshour	1-7
0500	RNZ-NAT	Checkpoint	2-6

Time	Station/Network	Program Name	Day(s)
0600	CRI-WASH	China Drive	1-7
1500	CRI-WASH	China Drive	1-7
2100	CRI-WASH	China Drive	1-7
1030	WRN-NA	Cmnwlth. B/C Assn: Pick of the Commonwealth	1/7
2300	ABC-RA	Connect Asia	1-5
1430	WRS-SUI	Drive Time	2-6
1530	RTE-R1	Drivetime	2-6
0330	PR-EXT	Europe East	7
0705	PR-EXT	Europe East	7
1205	PR-EXT	Europe East	1
1700	PR-EXT	Europe East	7
1930	PR-EXT	Europe East	7
1600	BBCWS-AM	Europe Today	2-6
1600	BBCWS-IE	Europe Today	2-6
1600	BBCWS-PR	Europe Today	2-6
1600	BBCWS-PR	Europe Today	2-6
2230	RTHK-3	Hong Kong Today	1-5
0030	DW	Inside Europe	7
0430	DW	Inside Europe	6
0630	DW	Inside Europe	6
0830	DW	Inside Europe	6
1030	DW	Inside Europe	6
1105	DW	Inside Europe	7
1230	DW	Inside Europe	6
1305	DW	Inside Europe	7
1430	DW	Inside Europe	6
1505	DW	Inside Europe	7
1630	DW	Inside Europe	6
1830	DW	Inside Europe	6
2030	DW	Inside Europe	6
2230	DW	Inside Europe	6
0705	DW	Inside Europe	7
1000	RNZ-NAT	Late Edition	2-6
1900	CBC-R1A	Local afternoon program	2-6
2100	CBC-R1C	Local afternoon program	2-6
2000	CBC-R1E	Local afternoon program	2-6
2200	CBC-R1M	Local afternoon program	2-6
2300	CBC-R1P	Local afternoon program	2-6
0845	CBC-R1A	Local morning program	2-6
1045	CBC-R1C	Local morning program	2-6
0945	CBC-R1E	Local morning program	2-6
1145	CBC-R1M	Local morning program	2-6
1245	CBC-R1P	Local morning program	2-6
1500	CBC-R1A	Local noon-hour program	2-6
1700	CBC-R1C	Local noon-hour program	2-6
1600	CBC-R1E	Local noon-hour program	2-6
1800	CBC-R1M	Local noon-hour program	2-6
1900	CBC-R1P	Local noon-hour program	2-6
1000	SABC-SAFM	Midday Live	2-6
0000	RNZ-NAT	Midday Report	2-6
0600	RTE-R1	Morning Ireland	2-6
1800	RNZ-NAT	Morning Report	1-5
1500	RNW	Network Europe	2-6
1600	RNW	Network Europe	2-6
2000	RNW	Network Europe	2-6
0805	DW	Network Europe Week	7
0905	DW	Network Europe Week	7
1005	DW	Network Europe Week	7
1205	DW	Network Europe Week	7
1405	DW	Network Europe Week	7
0330	PR-EXT	Network Europe Week	1/4
0705	PR-EXT	Network Europe Week	2
0000	CRI-ENG	News and Reports	1/7
0000	CRI-ENG	News and Reports	2-6
0100	CRI-ENG	News and Reports	1/7
0100	CRI-ENG	News and Reports	2-6
0300	CRI-ENG	News and Reports	1/7
0300	CRI-ENG	News and Reports	2-6
0400	CRI-ENG	News and Reports	1/7
0400	CRI-ENG	News and Reports	2-6
0500	CRI-ENG	News and Reports	1/7
0500	CRI-ENG	News and Reports	2-6
0600	CRI-ENG	News and Reports	1/7
0600	CRI-ENG	News and Reports	2-6

News Magazine Programs *continued*

Time	Station/Network	Program Name	Day(s)
1100	CRI-ENG	News and Reports	1/7
1300	CRI-ENG	News and Reports	1-5
1300	CRI-ENG	News and Reports	6/7
1400	CRI-ENG	News and Reports	1-5
1400	CRI-ENG	News and Reports	6/7
1500	CRI-ENG	News and Reports	1-5
1500	CRI-ENG	News and Reports	6/7
2300	CRI-ENG	News and Reports	6/7
0000	CRI-RTC	News and Reports	1-7
0100	CRI-RTC	News and Reports	1-7
0200	CRI-RTC	News and Reports	1-7
0300	CRI-RTC	News and Reports	1-7
0400	CRI-RTC	News and Reports	1-7
0500	CRI-RTC	News and Reports	1-7
0600	CRI-RTC	News and Reports	1-7
0700	CRI-RTC	News and Reports	1/7
0800	CRI-RTC	News and Reports	1-7
0900	CRI-RTC	News and Reports	1-7
1000	CRI-RTC	News and Reports	1/7
1100	CRI-RTC	News and Reports	1/7
1200	CRI-RTC	News and Reports	1-7
1300	CRI-RTC	News and Reports	1-7
1400	CRI-RTC	News and Reports	1-7
1500	CRI-RTC	News and Reports	1-7
1600	CRI-RTC	News and Reports	1-7
1700	CRI-RTC	News and Reports	1/7
1800	CRI-RTC	News and Reports	1-7
1900	CRI-RTC	News and Reports	1-7
2000	CRI-RTC	News and Reports	1-7
2100	CRI-RTC	News and Reports	1/7
2200	CRI-RTC	News and Reports	1/7
2300	CRI-RTC	News and Reports	1-7
1130	CRI-WASH	News and Reports	1-7
1230	CRI-WASH	News and Reports	1-7
1630	CRI-WASH	News and Reports	1-7
1200	RTE-R1	News At One	2-6
0200	RAE	News, Reports, Features, Tangos	3-7
1800	RAE	News, Reports, Features, Tangos	2-6
1200	BBCWS-AM	Newshour	1-7
2000	BBCWS-AM	Newshour	1-7
1200	BBCWS-IE	Newshour	1/7
1300	BBCWS-IE	Newshour	2-6
2000	BBCWS-IE	Newshour	1-7
1200	BBCWS-NX	Newshour	1/7
1200	BBCWS-NX	Newshour	2-6
2000	BBCWS-NX	Newshour	1-7
1200	BBCWS-PR	Newshour	1/7
1200	BBCWS-PR	Newshour	2-6
2000	BBCWS-PR	Newshour	1-7
0300	RNW	Newsline	3-7
0400	RNW	Newsline	3-7
0430	RNW	Newsline	3-7
0530	RNW	Newsline	3-7
0630	RNW	Newsline	3-7
0730	RNW	Newsline	3-7
0830	RNW	Newsline	3-7
1300	RNW	Newsline	2-6
1430	RNW	Newsline	2-6
1700	RNW	Newsline	2-6
1800	RNW	Newsline	2-6
1900	RNW	Newsline	2-6
2100	RNW	Newsline	2-6
2200	RNW	Newsline	2-6
0005	DW	Newslink	1-7
0105	DW	Newslink	1/2
0205	DW	Newslink	1/2
0305	DW	Newslink	2
0405	DW	Newslink	2
0405	DW	Newslink	3-7
0505	DW	Newslink	2
0505	DW	Newslink	3-7
0605	DW	Newslink	2
0605	DW	Newslink	3-7
0705	DW	Newslink	2-6

Time	Station/Network	Program Name	Day(s)
0805	DW	Newslink	2-6
0905	DW	Newslink	2-6
1005	DW	Newslink	2-6
1105	DW	Newslink	2-6
1205	DW	Newslink	2-6
1305	DW	Newslink	2-6
1405	DW	Newslink	2-6
1505	DW	Newslink	2-6
1605	DW	Newslink	1-7
1705	DW	Newslink	1-7
1805	DW	Newslink	1-7
1905	DW	Newslink	1-7
2005	DW	Newslink	1-7
2105	DW	Newslink	1-7
2205	DW	Newslink	1-7
2305	DW	Newslink	1-7
0105	DW	Newslink Plus	3-7
0205	DW	Newslink Plus	3-7
0305	DW	Newslink Plus	3-7
1000	RTHK-3	NewsWrap	2-6
0500	ABC-RA	Pacific Beat - Afternoon Edition	2-6
0700	ABC-RA	Pacific Beat - Afternoon Edition	2-6
1800	ABC-RA	Pacific Beat - Morning Edition	1-5
0300	ABC-RA	Pacific Review	7
0600	ABC-RA	Pacific Review	7
1800	ABC-RA	Pacific Review	6
2000	ABC-RA	Pacific Review	6
0800	ABC-RA	PM	2-6
1110	ABC-RA	PM	2-6
0700	ABC-RN	PM	2-6
1600	BBC-R4	PM	2-6
1600	BBC-R4	PM	7
1400	SABC-SAFM	PM Live	2-6
2000	ABC-RN	Radio National Breakfast	1-5
2130	ABC-RN	Radio National Breakfast	1-5
1930	WRN-NA	Radio New Zealand Int.: Dateline Pacific	1
1000	ORF-FM4	Reality Check (in German and English)	2-7
1100	CRI-WASH	Realtime China	1-7
1600	CRI-WASH	Realtime China	1-7
1800	WRN-NA	RTE Ireland: Drivetime	2-6
2100	WRN-NA	RTE Ireland: Drivetime	2-6
2110	ABC-RA	Saturday AM	6
2100	ABC-RN	Saturday AM	6
1800	SABC-SAFM	Saturday PM	7
0315	KBS-WR	Seoul Calling	2-6
0845	KBS-WR	Seoul Calling	2-6
1115	KBS-WR	Seoul Calling	2-6
1215	KBS-WR	Seoul Calling	2-6
2215	KBS-WR	Seoul Calling	2-6
2315	KBS-WR	Seoul Calling	2-6
1800	SABC-SAFM	Sunday PM	1
2200	CBC-R1A	The World This Weekend	1/7
2300	CBC-R1C	The World This Weekend	1/7
2200	CBC-R1E	The World This Weekend	1/7
0000	CBC-R1M	The World This Weekend	1/2
0100	CBC-R1P	The World This Weekend	1/2
0100	CBC-R1S	The World This Weekend	1/2
2200	CBC-R1S	The World This Weekend	1/7
0200	ABC-RA	The World Today	2-6
0200	ABC-RN	The World Today	2-6
0200	BBCWS-AM	The World Today	1/7
0200	BBCWS-AM	The World Today	2-6
0600	BBCWS-AM	The World Today	3-1
0700	BBCWS-AM	The World Today	1/7
0700	BBCWS-AM	The World Today	2-6
2300	BBCWS-AM	The World Today	2-6
0200	BBCWS-IE	The World Today	1-7
0300	BBCWS-IE	The World Today	1/7
0300	BBCWS-IE	The World Today	2-6
0500	BBCWS-IE	The World Today	2-6
0600	BBCWS-IE	The World Today	1/7
0600	BBCWS-IE	The World Today	2-6
0700	BBCWS-IE	The World Today	1/7
0700	BBCWS-IE	The World Today	2-6
0200	BBCWS-NX	The World Today	1/7

Time	Station/Network	Program Name	Day(s)
0200	BBCWS-NX	The World Today	2-6
0300	BBCWS-NX	The World Today	1/7
0500	BBCWS-NX	The World Today	1/7
0500	BBCWS-NX	The World Today	2-6
0700	BBCWS-NX	The World Today	1/7
2300	BBCWS-NX	The World Today	1-5
2300	BBCWS-NX	The World Today	6/7
0200	BBCWS-PR	The World Today	1/7
0200	BBCWS-PR	The World Today	2-6
0300	BBCWS-PR	The World Today	1/7
0500	BBCWS-PR	The World Today	1
0500	BBCWS-PR	The World Today	2-6
0500	BBCWS-PR	The World Today	7
0700	BBCWS-PR	The World Today	1/7
2300	BBCWS-PR	The World Today	1-5
2300	BBCWS-PR	The World Today	6/7
2100	BBC-R4	The World Tonight	2-6
0500	BBC-R4	Today	2-6
0600	BBC-R4	Today	7
0000	RTHK-3	Today at Eight	7
2300	RTHK-3	Today at Seven	6
0440	SABC-CHAF	UN Chronicle	2
0540	SABC-CHAF	UN Chronicle	2
0615	SABC-CHAF	UN Chronicle	2
1015	SABC-CHAF	UN Chronicle	2
0045	WRN-NA	UN Radio: The UN and Africa	2
1045	WRN-NA	UN Radio: The UN and Africa	1/7
1600	WRN-NA	UN Radio: UN Calling Asia	1
0000	WRN-NA	UN Radio: UN Today	3-7
0845	WRN-NA	UN Radio: Women/Perspective	7
0010	NHK-RJ	What's Up Japan	3/5/7
0010	NHK-RJ	What's Up Japan	4/6
0510	NHK-RJ	What's Up Japan	2/4/6
0510	NHK-RJ	What's Up Japan	3/5
1210	NHK-RJ	What's Up Japan	2/4/6
1210	NHK-RJ	What's Up Japan	3/5
1410	NHK-RJ	What's Up Japan	2/4/6
1410	NHK-RJ	What's Up Japan	3/5
0710	RTE-R1	World Report	7
0900	BBCWS-AM	World Update	2-6
0900	BBCWS-NX	World Update	2-6
0900	BBCWS-PR	World Update	2-6

Personal Interview Programs

Time	Station/Network	Program Name	Day(s)
0810	RTE-R1	Conversations with Eamon Dunphy	7
0400	PR-EXT	Day in the Life	5
0730	PR-EXT	Day in the Life	5
1205	PR-EXT	Day in the Life	3
1720	PR-EXT	Day in the Life	4
1950	PR-EXT	Day in the Life	4
0430	VOR-WS	Guest Speaker	3-7
0800	WRS-SUI	Interview of the Week	1
0240	RTI	On the Line	1
0110	R.PRG	One on One	3
0119	R.PRG	One on One	1
0210	R.PRG	One on One	3
0219	R.PRG	One on One	1
0410	R.PRG	One on One	3
0419	R.PRG	One on One	1
0440	R.PRG	One on One	3
0449	R.PRG	One on One	1
0810	R.PRG	One on One	2
0819	R.PRG	One on One	7
1140	R.PRG	One on One	2
1149	R.PRG	One on One	7
1410	R.PRG	One on One	2
1419	R.PRG	One on One	7
1710	R.PRG	One on One	2
1719	R.PRG	One on One	7
1810	R.PRG	One on One	2

Personal Interview Programs *continued*

Time	Station/Network	Program Name	Day(s)
1819	R.PRG	One on One	7
2110	R.PRG	One on One	2
2119	R.PRG	One on One	7
2240	R.PRG	One on One	2
2249	R.PRG	One on One	7
2340	R.PRG	One on One	2
2349	R.PRG	One on One	7
1300	WRN-NA	RTE Ireland: Conversations with Eamon Dunphy	7
0230	BBCWS-AM	The Interview	1
0330	BBCWS-AM	The Interview	1
1330	BBCWS-AM	The Interview	1
2130	BBCWS-AM	The Interview	7
0330	BBCWS-IE	The Interview	1
0630	BBCWS-IE	The Interview	1
1130	BBCWS-IE	The Interview	7
1730	BBCWS-IE	The Interview	1
2130	BBCWS-IE	The Interview	7
0030	BBCWS-NX	The Interview	1
0230	BBCWS-NX	The Interview	1
0330	BBCWS-NX	The Interview	7
0630	BBCWS-NX	The Interview	1
1130	BBCWS-NX	The Interview	1
1330	BBCWS-NX	The Interview	7
1730	BBCWS-NX	The Interview	1
1930	BBCWS-NX	The Interview	7
2130	BBCWS-NX	The Interview	7
2330	BBCWS-NX	The Interview	6
0330	BBCWS-PR	The Interview	7
0730	BBCWS-PR	The Interview	4
1130	BBCWS-PR	The Interview	7
1330	BBCWS-PR	The Interview	7
1730	BBCWS-PR	The Interview	1
1830	BBCWS-PR	The Interview	4
1930	BBCWS-PR	The Interview	7
2130	BBCWS-PR	The Interview	7
2330	BBCWS-PR	The Interview	6
0730	RFI	Voices	4
1430	RFI	Voices	4

Press Review Programs

Time	Station/Network	Program Name	Day(s)
0355	PR-EXT	A Look at the Weeklies	1
0725	PR-EXT	A Look at the Weeklies	1
1225	PR-EXT	A Look at the Weeklies	7
1720	PR-EXT	A Look at the Weeklies	7
1950	PR-EXT	A Look at the Weeklies	7
0415	RFI	French Press Review	2-6
0515	RFI	French Press Review	2-6
0715	RFI	French Press Review	2-6
2050	RNZI	New Zealand Newspaper Headlines	1-5
0700	RTE-R1	News, Sports, It Says in the Papers	1/7
0800	RTE-R1	News, Sports, It Says in the Papers	1/7
0520	WRS-SUI	Swiss Press Review	2-6
0630	WRS-SUI	Switzerland Today	2-6
0620	WRS-SUI	World Press Review	2-6

Sports Programming

Time	Station/Network	Program Name	Day(s)
1730	RTE-R1	Drivetime Sport	2-6
1630	SABC-SAFM	Gameplan	2-6
0325	VOR-WS	Legends of Russian Sports	2
0340	VOR-WS	Legends of Russian Sports	1/4/7
0840	VOR-WS	Legends of Russian Sports	1/4
0945	VOR-WS	Legends of Russian Sports	2
2200	RTE-R1	News and GAA Sports Results	1
0400	PR-EXT	Offside	7
0725	PR-EXT	Offside	7
1230	PR-EXT	Offside	6
2100	WRN-NA	RTE Ireland: Sport	2-6

Sports Programming *continued*

Time	Station/Network	Program Name	Day(s)
1300	SABC-SAFM	SAfm Sports Special	7
1425	RFI	Sport	2-7
0315	SABC-CHAF	Sport	7
0415	SABC-CHAF	Sport	7
1030	ABC-RA	Sports Factor (til 1/09)	6
1530	ABC-RA	Sports Factor (til 1/09)	6
0120	R.PRG	Sports News	3
0220	R.PRG	Sports News	3
0420	R.PRG	Sports News	3
0450	R.PRG	Sports News	3
0820	R.PRG	Sports News	2
1150	R.PRG	Sports News	2
1420	R.PRG	Sports News	2
1720	R.PRG	Sports News	2
1820	R.PRG	Sports News	2
2120	R.PRG	Sports News	2
2250	R.PRG	Sports News	2
2350	R.PRG	Sports News	2
2350	R.PRG	Sports News	2
1810	RNZI	Sports News	1-6
2010	RNZI	Sports News	1-6
2110	RNZI	Sports News	1-5
2210	RNZI	Sports News	1-5
0300	ABC-RA	Sports Report	2-6
0530	ABC-RA	Sports Report	2/3
0600	ABC-RA	Sports Report	2-6
0730	ABC-RA	Sports Report	2-6
1100	ABC-RA	Sports Report	2-6
1550	BBCWS-AM	Sports Report	2-6
0420	BBCWS-NX	Sports Report	1/7
0030	DW	Sports Report	1/2
0130	DW	Sports Report	1/2
0330	DW	Sports Report	1/2
0415	DW	Sports Report	2
0430	DW	Sports Report	1
0515	DW	Sports Report	2
0615	DW	Sports Report	2
0630	DW	Sports Report	1
1930	DW	Sports Report	1/7
2130	DW	Sports Report	1/7
2330	DW	Sports Report	1/7
0150	BBCWS-AM	Sports Roundup	3-6
0420	BBCWS-AM	Sports Roundup	1/7
0450	BBCWS-AM	Sports Roundup	2-6
1020	BBCWS-AM	Sports Roundup	1/7
1050	BBCWS-AM	Sports Roundup	2-6
2150	BBCWS-AM	Sports Roundup	2-6
0020	BBCWS-IE	Sports Roundup	1/7
0050	BBCWS-IE	Sports Roundup	2-6
0420	BBCWS-IE	Sports Roundup	1/7
0450	BBCWS-IE	Sports Roundup	2-6
0920	BBCWS-IE	Sports Roundup	1/7
1050	BBCWS-IE	Sports Roundup	2-6
2120	BBCWS-IE	Sports Roundup	1
0020	BBCWS-NX	Sports Roundup	1/7
0050	BBCWS-NX	Sports Roundup	2-6
0150	BBCWS-NX	Sports Roundup	2-6
0450	BBCWS-NX	Sports Roundup	2-6
0850	BBCWS-NX	Sports Roundup	2-6
1020	BBCWS-NX	Sports Roundup	1/7
1050	BBCWS-NX	Sports Roundup	2-6
1150	BBCWS-NX	Sports Roundup	2-6
1450	BBCWS-NX	Sports Roundup	2-6
1550	BBCWS-NX	Sports Roundup	2-6
1850	BBCWS-NX	Sports Roundup	2-6
1950	BBCWS-NX	Sports Roundup	2-6
2120	BBCWS-NX	Sports Roundup	1
2150	BBCWS-NX	Sports Roundup	2-6
2250	BBCWS-NX	Sports Roundup	1-5
1400	BBCWS-AM	Sportsworld	7
1400	BBCWS-IE	Sportsworld	7
1400	BBCWS-NX	Sportsworld	7
1500	BBCWS-NX	Sportsworld	1
1400	BBCWS-PR	Sportsworld	7

Sports Programming *continued*

Time	Station/Network	Program Name	Day(s)
2000	RNZI	Sportsworld	7
1730	BBCWS-AM	Sportsworld Extra	7
1830	BBCWS-AM	Sportsworld Extra	1
1730	BBCWS-IE	Sportsworld Extra	7
1830	BBCWS-IE	Sportsworld Extra	1
1730	BBCWS-NX	Sportsworld Extra	7
1830	BBCWS-NX	Sportsworld Extra	1
1300	RTE-R1	Sunday Sport with Adrian Eames	1
1500	BBCWS-IE	Sunday Sportsworld	1
1630	CBC-R1A	The Inside Track	1
1800	CBC-R1A	The Inside Track	4
1830	CBC-R1C	The Inside Track	1
2000	CBC-R1C	The Inside Track	4
1730	CBC-R1E	The Inside Track	1
1900	CBC-R1E	The Inside Track	4
1930	CBC-R1M	The Inside Track	1
2100	CBC-R1M	The Inside Track	4
2200	CBC-R1P	The Inside Track	4
2330	CBC-R1P	The Inside Track	1
2230	CBC-R1S	The Inside Track	7
0900	ABC-RN	The Sports Factor (til 1/09)	6
1530	ABC-RN	The Sports Factor (til 1/09)	6
2230	ABC-RN	The Sports Factor (til 1/09)	5
0100	ABC-RA	Total Rugby	7
0700	ABC-RA	Total Rugby	7
1700	ABC-RA	Total Rugby	1
0130	BBCWS-AM	World Football	7
1130	BBCWS-AM	World Football	7
0330	BBCWS-IE	World Football	7
0630	BBCWS-IE	World Football	7
0930	BBCWS-IE	World Football	7
0330	RNZI	World in Sport	5
0730	RNZI	World in Sport	5
1130	RNZI	World in Sport	4
1330	RNZI	World in Sport	4
1708	RNZI	World in Sport	4
2115	RNZI	World in Sport	4

Science & Technology Programs

Time	Station/Network	Program Name	Day(s)
0330	BBCWS-AM	Discovery	5
0830	BBCWS-AM	Discovery	4
1330	BBCWS-AM	Discovery	4
1430	BBCWS-AM	Discovery	4
2230	BBCWS-AM	Discovery	4
0230	BBCWS-IE	Discovery	1
1230	BBCWS-IE	Discovery	4
1530	BBCWS-IE	Discovery	4
1930	BBCWS-IE	Discovery	4
2330	BBCWS-IE	Discovery	4
1230	RNZ-NAT	Discovery	1
1900	SABC-SAFM	Earth Hour	5
1400	ABC-RN	Future Report (from 2/09)	5
2230	ABC-RN	Future Report (from 2/09)	4
1030	ABC-RA	Futures Report (from 2/09)	5
1530	ABC-RA	Futures Report (til 2/09)	5
1835	ABC-RN	In Conversation	5
0935	ABC-RN	In Conversation (til 1/09)	5
0430	ABC-RA	Innovations	7
0830	ABC-RA	Innovations	1
1330	ABC-RA	Innovations	2
1700	ABC-RA	Innovations	2
2230	ABC-RA	Innovations	7
2000	BBC-R4	Leading Edge	5
0345	ABC-RA	Ockham's Razor	7
0645	ABC-RA	Ockham's Razor	7
0710	ABC-RA	Ockham's Razor	1
2245	ABC-RN	Ockham's Razor	7
0906	RNZ-NAT	Our Changing World	5
1315	RNZ-NAT	Our Changing World	7
0200	CBC-R1A	Quirks and Quarks	3

Science & Technology Programs *continued*

Time	Station/Network	Program Name	Day(s)
1500	CBC-R1A	Quirks and Quarks	7
0400	CBC-R1C	Quirks and Quarks	3
1700	CBC-R1C	Quirks and Quarks	7
0300	CBC-R1E	Quirks and Quarks	3
1600	CBC-R1E	Quirks and Quarks	7
0500	CBC-R1M	Quirks and Quarks	3
1800	CBC-R1M	Quirks and Quarks	7
0600	CBC-R1P	Quirks and Quarks	3
1900	CBC-R1P	Quirks and Quarks	7
1200	CBC-R1S	Quirks and Quarks	7
1900	CBC-R1S	Quirks and Quarks	7
0330	BBCWS-AM	Science in Action	7
0830	BBCWS-AM	Science in Action	7
1330	BBCWS-AM	Science in Action	6
1430	BBCWS-AM	Science in Action	6
2230	BBCWS-AM	Science in Action	6
0130	BBCWS-IE	Science in Action	2
1230	BBCWS-IE	Science in Action	6
1530	BBCWS-IE	Science in Action	6
1930	BBCWS-IE	Science in Action	6
2330	BBCWS-IE	Science in Action	6
1030	BBCWS-PR	Science in Action	1
1530	BBC-R4	Science or health series or feature	4
2000	BBC-R4	Science or health series or feature	2/3
0400	VOR-WS	Science Plus	5
0500	VOR-WS	Science Plus	4
0700	VOR-WS	Science Plus	2
0800	VOR-WS	Science Plus	1
0030	DW	Spectrum	4
0430	DW	Spectrum	3
0630	DW	Spectrum	3
0830	DW	Spectrum	3
1030	DW	Spectrum	3
1230	DW	Spectrum	3
1430	DW	Spectrum	3
1630	DW	Spectrum	3
1830	DW	Spectrum	3
2030	DW	Spectrum	3
2230	DW	Spectrum	3
1530	BBC-R4	The Material World	5
1600	ABC-RA	The Science Show	1
0200	ABC-RN	The Science Show	7
0900	ABC-RN	The Science Show	2
1800	ABC-RN	The Science Show	2
1000	BBC-R4	World on the Move	3
2000	BBC-R4	World on the Move	4

Program Name and Description List

Program Name	Description
A Book at Bedtime	Modern classics, new works by leading writers and literature from around the world.
A Good Read	Sue MacGregor and guests talk about their favorite books.
A Propos	Jim Corcoran plays the best recordings from Francophone Canada with special emphasis on the Quebec popular music scene.
A Stroll Around the Kremlin	The past and present of the heart of Russia's capital.
A World of Music	A program of the world's diverse music ranging from the traditional to the cutting edge, produced by Rick Fulker.
Afghanada	Dramatic series that probes the war in Afghanistan through the eyes of Canadian soldiers.
Africa Rise and Shine	A morning newsmagazine from South Africa for the African continent.
Africa This Week	A roundup of Channel Africa's top stories of the week.
African Connection	Richard Nwamba with the best selection of music from all across Africa and the islands.
Afternoon Talk	Commentary, humor and issue debates on politics, arts, culture and lifestyle with Karabo Kgoleng.
Airplay	Jane Ulman presents a weekly program of new Australian radio writing and performance.
All in the Mind	The mental "universe": the mind, brain and behavior.
All the Way with Ray	RTHK's longest running show, "Uncle" Ray Cordeiro presents a relaxed program packed with all your old favorites.
AM	Australian national current affairs.
AM Live	Tsepiso Makwetia with the morning news, insights, analysis and debate in and from South Africa.
And the Winner Is	An encore of award winning CBC Radio and international documentaries.
Any Questions? Any Answers?	A forum for lively debate between leading figures chaired by Jonathan Dimbleby.
Around Poland	Exploring the Poland outside the cities.
Arts on the Air	Breandain O'Shea presents a weekly magazine highlighting European artists and cultural events.
Artworks	The big themes, views, issues and events in the arts in Australia and overseas.
As It Happens	The current affairs program that phones out to where the news is happening.
Asia Pacific	A daily current affairs program covering Asia and the Pacific.
Asia Review	The week's current events in the region.
Assignment	BBC correspondents report on events and themes in the news.
At the Movies	A weekly topical magazine about current film releases and film related topics.
Australia All Over	An eclectic mix of music, poetry, anecdotes, book readings and audience interaction hosted by "Macca" a/k/a Ian McNamara.
Australia Talks	A daily national phone-in program.
Australian Country Style	John Nutting hosts this popular weekly show highlighting country music and its artists.
Australian Express	All about Australia and what makes it tick, hosted by Roger Broadbent.
AWAYE!	Australian indigenous peoples' issues, culture and arts.
Backchat	Hong Kong residents have their say on the issues of the day moderated by Hugh Chiverton.
Background Briefing	Award-winning Australian investigative journalism.
Beale Street Caravan	Live performances drawn from concerts recorded in the U.S. for The Blues Foundation.
Between the Covers	Novels and short stories read in quarter hour installments with emphasis on Canadian contemporary fiction.
Beyond Beijing	A freewheeling program of Chinese and international popular music with cultural, lifestyle and entertainment features.
Big Ideas	Lectures,, conversations, features and special series from Australia and around the world.
Biz China	Business and financial news, with primary focus on the Chinese market.
Book Club	Readings from contemporary literature.
Book of the Week	Non-fiction, biography, autobiography, travel, diaries, essays, humor and history.
Book Reading (ABC)	Broadcast since 1948, the best of classic and contemporary fiction by Australian and world writers, read by some of Australia's finest actors.
Bookmark	Three Swiss writers choose and discuss a favorite book in a monthly program with Pete Forster
Breakfast Club	A morning program for Asia and the Pacific.
Bridges with Africa	A mix of lively discussion and thought-provoking reports linking African diaspora groups in Europe with their home countries and African radio stations.
Broadcasting House	The week's news with Paddy O'Connell.
Bush Telegraph	Michael Mackenzie with an entertaining look at rural and regional issues around Australia.
Business Daily	Issues and trends emerging in the global business market.
Business Week	Economic development in Poland.
By Design	Alan Saunders looks at how the way things look, feel and function reinvent the world to keep pace with changing needs and desires.
C'est la vie	Life in French-speaking Canada explored through interviews and documentaries, with Bernard St-Laurent.
CBC Overnight	Information programs from public broadcasters around the world.
Charlie Gillett's World of Music	He plays the best of world music.
Chart Show	The top selling music in Poland.
Checkpoint	Mary Wilson presents the day's major international and New Zealand stories.
China Beat	New, contemporary and uniquely Chinese music.
China Drive	A bilingual (English and Chinese) daily drivetime features and music magazine with lifestyle, entertainment and practical news you can use.
China Horizons	A magazine program focusing on the rapid economic and social developments in China's provinces, the natural and historic features of the ancient country and the diversified customs and culture of its 56 ethnic groups.
Classic Serial	Dramatizations of works which have achieved--or are on their way to achieving--classic status.
Classical Sunday	A full symphony in the first half, pops the second with Simon Lomberg in Johannesburg and John Orr in Capetown.
Club 9516	David Page's tour-de-force of music, listener letters, interviews and commentary.
Concert Hour	Musical performances from Germany's palaces, festivals, churches and concert halls.
Connect Asia	News, views and analysis on the stories that matter in Asia.
Cool	The latest in youth culture in Germany and across Europe with Anke Rasper
Cool Trax	Late night sounds with the accent on rhythm and blues and soul with Francis Chan.
Correspindents Report	ABC (Australia) overseas reporters interpret and analyze the week's major events.
Correspondent's Notebook	A personal, professional eyewitness perspective on a major news story or issue in the Asia Pacific region.
Counterpoint	Challenging commentary on a range of Australian social, economic and cultural issues.
Country Breakfast	A weekly look at Australian rural and regional issues.
Country Life	A weekly program of issues and stories from New Zealand rural communities.
CRI Roundup	A weekly digest of the major news stories from China and around the world.
Cross Country Checkup	Rex Murphy presides over Canada's only national open line radio program.

Program Name	Description
Crossroads	Relatiions between the French and African communities.
Culture Shock	The latest global trends and ideas driving human behavior.
Curious Orange	Michael Walraven and Ashleigh Elson present a weekly guide to modern day Holland.
Czech Books	Czech writers and literature today.
Czechs Today	Personalities shaping contemporary Czech society.
Dateline Pacific	Daily roundup of the latest news from the Pacific region.
Day in the Lifc	Interviews with Poles from the prominent to everyday people.
Deep Roots	Tom Power plays the best folk, traditional and roots music from Canada and around the world.
Definitely Not the Opera	Sook-Yin Lee explores the nooks and crannies of popular culture in Canada and the world.
Desert Island Discs	Kirsty Young invites a guest to choose the eight records they would take with them to a desert island.
Dialogue	Reports, background and insights on religious, social and cultural movements around the world.
Dig It	Hester Macdonald's horticultural hints for growing things in Switzerland.
Digital Planet	The weekly BBC World Service technology program with Gareth Mitchell and Bill Thompson.
Discovery	A weekly exploration of today's most significant scientific discoveries.
Dispatches	Rich Macinnes-Rae with reports and documentaries focusing on covering global issues and international current events.
Drive Time	A fast, fun and furious mix of gossip, news and music plus the buzz on what's happening across Switzerland.
Earth Hour	Earth science and environment with the latest breakthroughs in local and international science.
Earthbeat (on RNW)	Examining what we grow, build, consume and destroy; and how that cycle affects the planet.
Ed Reardon's Week	A serial whose episodes track Ed's flawed attempts to escape poverty and gain literary success.
Encounter	Florence Spurling with a highly acclaimed series that examines the connections between religion and life with a special emphasis on the religious experience of multicultural Australia.
Encyclopedia "All Russia"	A history of Russian civilization based on the Russian alphabet.
Europe East	Reports from a network of correspondents from all over the region.
Europe Today	A slice of life from European cities through the eyes of the BBC's contributors, looking at trends and developments across the region.
Eurovox	Kateri Jochum with the latest news and views about European lifestyles, attitudes, design and fashion
Faith to Faith	A weekly religious panel discussion mediated by Peter-James Smith.
Farming Today	The news for those who live, work or have an interest in the countryside.
Feedback	Roger Bolton hears listener reactions to Radio 4 programming and policies.
Festival of Funny	Stand-up comedy from Canada and around the world.
File on 4	Current affairs documentaries.
First Person	A serialized reading of a published autobiography.
Focus	The arts in Poland.
Focus on Asia and the Pacific	News and comment on events in the region.
Focus on Politics	A weekly analysis of significant political issues in New Zealand.
Folk Box	The traditional music of the Russian Commonwealth's many ethnic groups.
FORA Radio	Talks, lectures and debates from Australia and around the world.
From Fact to Fiction	Writers create a fictional response to the week's news.
From Our Own Correspondent	BBC correspondents share their perceptions of life in their part of the world.
From the Archives	Sampling some of the unique recordings in the archives of Czech Radio.
Front Row	Radio 4's live magazine program on the world of arts, literature, film, media and music.
Frontline	The stories behind the controversial and sometimes difficult legal cases encountered by ordinary Chinese citizens.
Future Report	Current world trends on globalization, communications technologies and the shifting cultural, social, political and economic responses to them.
Gadget Guru	The latest in consumer technology and high-tach developments.
Gameplan	Kwena Moabelo with a sports magazine show.
Garrison Keillor's Radio Show	A version of "A Prairie Home Companion" edited for international listeners.
Gateway to Africa	Reports on South African economic and financial affairs.
Give Me Five	Hong Kong and international personalities pick five pieces of music that have influenced their lives.
Global Arts and Entertainment	A weekly compilation of the best from the BBC World Service's arts and culture programs.
Global Business	Peter Day explores the forces and issues driving the world of business.
Go 4 It!	A children's program.
Go!	Brent Bambury with a live Canadian Saturday morning show of music, humor, talk and performances.
Guest Speaker	Vitaly Glazunov talks with leading Russian and foreign political scientists, economists and artists.
Health Check	The issues affecting the world of medicine and healthcare.
Health Matters	Monica Morrell with new perspectives on mental and physical health issues.
Health Report	Making medicine understandable.
Heart and Soul	An exploration of religious belief and how it affects people's lives.
Hidden Treasures	Musical gems from niche markets around the globe.
High Note	Classical music from Poland.
Hindsight	Australian social history.
Hits in Germany	Deborah Freedman brings you the latest sounds, interviews with artists and news from the vibrant German music scene.
Hong Kong Heritage	Anne Marie Evans explores Hong Kong and its social, cultural, architectural and artistic heritage.
Hong Kong Today	Nick Beaucroft and Bryan Curtis with local and international news and current affairs, finance and business news, sports and studio guests.
Ideas (CBC)	Talks and documentaries covering social issues, culture and the arts, geopolitics, history, science and technology, biology and the humanities.
Ideas (RNZ)	A weekly program exploring a range of philosophical, social, historical and environmental ideas.
In Business	Peter Day looks at new ways of work and new technologies in the world of business.
In Conversation	Robyn Williams talks both to scientists and those interested in the subject.
In Our Time	Melvyn Bragg and guests investigate the history of ideas.
In the Key of Charles	Gregory Charles shares his love of music from classical to jazz to pop in an entertainingly eclectic program.
In the Loop	A lively, varietal mix of music, talk and sounds from Oceania.
In the Spotlight	A culture journey of China with its variety of ethnic cultures.
In Touch	News, views and information for people who are blind or partially sighted.
In Touch	Listeners comment about Radio Polonia programs.
In-Box	Margot Forbes and Rita Oliver read and respond to listener letters and questions to Deutsche Welle.
Innovations	A showcase of Australian design, discovery, invention, engineering and research.
Inside Europe	A weekly news magazine exploring the topical issues that shape the continent co-produced by Helen Seeney and Barbara Gruber.
Inside the Music	Patti Schmidt explores the whys and wherefores of music through documentaries, interviews and series.

Program Name	Description
Insight	Perspectives on events and issues shaping the globe.
Inspector Steine	A comedic drama series by Lynne Truss, set in 1950s Brighton, England.
Inspired Minds	Breandain O'Shea interviews the world's most talented artists as they tour Germany and Europe.
Into the Music	Exploring and celebrating all aspects of music with Robyn Johnston.
Intune	Sibahle Malinga educates and informs young people on issues relating to their lives.
iPM	A weekly radio program, blog and podcast by Rupert Allman stressing audience interaction, ideas, contribution and production.
Jade Bells and Bamboo Pipes	Authentic traditional, classical Chinese music with Carlson Wong.
Jazz Notes	Australian jazz
Jazz Show	Russian jazz performances and performers.
Kaleidoscope	The latest economic, social and cultural events in the Russian Commonwealth; and the traditions and customs of its peoples.
Kids in Mind	Rachel Melville-Thomas takes listener questions about raising children.
Last Word	Radio 4's weekly obituaries program.
Late Edition	The most significant news stories and interviews of the day in New Zealand.
Late Night Live	Conversations about politics, current events, philosophy and culture hosted by Phillip Adams.
Laugh Out Loud	Sabrina Jalees features the best and funniest comics in the business.
Law Report	All about the law makers and the law breakers.
Law Report (SAfm)	John Orr tackles the law in South Africa.
Leading Edge	Geoff Watts brings you the latest news, controversies and conversations from the world of science.
Legends of Russian Sports	Profiles of prominent Russian athletes.
Letter from Poland	An irreverent look at Poland by Londoner, Peter Gentle.
Letter from Prague	A personal view of life in and around the Czech capital.
Letter to Hong Kong	Leaders from Hong Kong's political parties and government departments take their turn to have their say.
Life and Times	Biographies of past figures from politics, literature, science, the arts, religion and all areas of public life.
Life Matters	A unique, daily interview program about social change and day-to-day life in Australia.
Lingua Franca	Maria Zjilstra with interviews and talks about all aspects of language.
Listener's Corner	A weekend magazine answering listener questions about China, insights into Chinese culture, history and humor, and a short Chinese language lesson.
Living Planet	News, background reports, interviews and features on environmental matters around the world.
Living Sounds	International and local gospel music with Khanyi Magubane.
Magazine	The show that starts where the news ends--the stories from the Czech Republic that you might otherwise have missed.
Magic Carpet	Czech world music.
Mailbox	A fortnightly review of listener mail and DX news with Myra Oh and Adrian Sainsbury.
Mailbox	Radio Prague replies to your letters, e-mails and phone calls and answers your questions.
Making History	Vanessa Collingridge and her team answer listeners' historical queries and celebrate the way in which we all make history.
Market Report	The world of high finance, share trading, mergers, acquisitions and the economy in South Africa.
Masala Canada	Wojtek Gwiazda with an eclectic mix of stories, conversations and music with a South Asian flavor.
Media Report	About the media industry and its future.
Media@SAfm	A weekly look at all South African media matters--advertising, marketing, public relations and branding.
Mediawatch	A critical look at the New Zealand media.
Midday Live	Lunch time news, business reports with informed analysis and comment from the newsmakers themselves.
Midday Report	World, New Zealand, business, rural and sports news with weather reports.
Midweek	Lively and diverse conversation with Libby Purves.
Mission Europe	An interactive radio and online language course teaching German, French and Polish.
Mission Paris	A dramatized language lesson in beginning French.
Money Box	Financial advice.
Money Talks	The latest German and European business trends and financial developments with news, reports and interviews.
Morning Report	Authoritative and comprehensive coverage of New Zealand and world events.
Morning Talk	One of South Africa's top talk shows tackling issues like crime, unemployment, politics, personal finance and health.
Moscow Mailbag	A question and answer show based on listener letters.
Moscow Yesterday and Today	The history of Russia's capital city.
MovieTime	A lively, entertaining and comprehensive review of movies, with interviews and behind-the-scenes information.
Multimedia	News, chat and interviews for those passionate about radio, hosted by Slawek Szefs and Marek Lasota.
Music 101	The best of New Zealand music as well as a Kiwi perspective on great music from the rest of the world.
Music and Musicians	Classical music performances by world famous composers and artists.
Music Calendar	Musicians and music events that have stood the test of time.
Music Deli	Paul Petran presents a wide variety of music styles recorded at live venues throughout Australia.
Music Profile	Introducing Czech musicians--groups, solo artists past and present.
Musical Tales	Information, little known facts, popular music and rare recordings illuminating the past and present of Russian music.
Naked Lunch	Sarah Passmore celebrates Hong Kong and the people who make it "buzz,", plus new music and classic tracks.
Neil Chase in New York	A weekly mix of music and comment from the Big Apple.
NEPAD Focus	A program highlighting the work of the New Partnership for Africa's Development.
Network Europe	Daily news, current affairs and culture from Europe's widest partnership of international broadcasters.
Network Europe Extra	Arts and culture from Europe's widest partnership of international broadcasters.
Network Europe Week	A digest of top stories from the daily Network Europe program.
New Music Releases	A sample of the latest Kiwi music hosted by Hana Tatere.
News and Reports	CRI's flagship news magazine featuring headlines, reports on major issues, regional coverage, press clippings, stocks, science and technology news and the weather.
News and Views	Russian views on news developments.
Newshour	News and analysis from around the globe.
Newsline	News and background with interviews and reports.
Newslink	Daily in-depth coverage and analysis of European and global current affairs
Newslink Plus	Breaking news, international current affairs and reports from Germany and around Europe
Nine to Noon	Everything from hard news to lifestyle features.
Ockhan's Razor	Thoughtful people have their say about issues in science.
Offside	Coverage of the Polish and European sports scene.
On the Beat	Peter Forster guides you through complex issues in the news and the important topics of the day.
On Your Farm	Getting to the heart of country life.

Program Name	Description
One in Five	The issues and experience of disability.
One on One	An informal interview of a Czech personality.
One Planet	A weekly view of global development and its impact on the environment.
Open Air	Polish society, entertainment, history and culture.
Open Book	Spotlighting new fiction and non-fiction, picking out the best of the paperbacks, talking to authors and publishers, and unearthing lost masterpieces.
Open Country	Local people making their corner of rural Britain unique.
Otherwise	Life in South Africa from the perspective of 52% of the population – women.
Our Changing World	Coverage of all scientific disciplines, natural history, environmental issues and health.
Our Heritage	Considerations on what it means to be African today.
Outfront	Stories, experimental audio and new ways of making radio that highlight Canadian perspectives about the Canadian experience.
Outlook	Human interest stories behind the headlines.
Over to You	Listeners comment on BBC World Service programs.
Pacific Beat	News and current affairs about the Pacific.
Pacific Beat - On the Mat	Issues of regional interest are discussed.
Pacific Correspondent	RNZI regional correspondents talk to Ben Lowings about political and social issues in their respective Pacific countries.
Pacific Review	A weekly round-up of the major stories from the region.
Panorama	A weekly foray into all things Czech, from cultural and artistic trends to social phenomena.
People in the Know	Political current affairs in China as seen by leading Chinese and international public figures.
Perspective	Opinion makers from Australia and overseas talk about issues which affect us all.
Phil Jupitus' Strips	A program about comic strips and animation.
Pick of the Week	Peter White presents a selection of highlights from the past week on BBC radio.
PM	A comprehensive evening current affairs round-up from Australia.
PM Live	A complete wrap of the day's news featuring analysis, business, sport and entertainment news.
PoeticA	The performance of poetry hosted my Mike Ladd.
Poetry Please	Listeners request poems to be read.
Politics UK	A roundup of the week's events in British politics.
Pop Up Japan	Japanese pop music
Q	Jian Ghomeshi with Canada's daily arts, culture and entertainment magazine.
Questions, Questions	Radio 4's listener-led problem solving program with Stewart Henderson.
Quiet Space	Ambient music incorporating 20th century composition, pop and ethereal tones.
Quirks and Quarks	The CBC's long-running (25 years +) award-winning, accessible science program hosted by Bob McDonald.
Radio Books	A selection of short stories by Dutch and Flemish writers in English translation.
Radio D	An interactive radio and online German language course for beginners.
Radio Eye	Brent Clough with the best features and documentaries from Australia and around the world.
Radio National Breakfast	Fran Kelly with comprehensive coverage and analysis of national and international events.
Ramblings	Clare Balding explores favorite British walks.
Realtime China	A news magazine looking at Chinese politics, society, economics and culture.
Rear Vision	A look at current issues from where we've been, where we are today and where we might be going.
Reflections from Asia	Leading Hong Kong journalist Harvey Stockwin with an award winning weekly in-depth analysis of Asian issues.
Religion Report	Religious affairs in Australia and around the world.
Reloaded	Radio Netherlands Worldwide's weekly highlights program presented by Mindy Ran.
Rendezvous	Life in the regions of France.
Reporting Religion	Reports on religious events and issues around the world.
Reports from Developing Countries	A focus on development efforts in Asia, Africa and Latin America.
Rewind	The best from the CBC's radio archives.
Rural Reporter	Stories from the Australian bush.
Russia and the World	Russia's relations with the international community.
Russia—1000 Years of Music	Year by year, century by century through the history of Russian music.
Russian Hits	Russia's top pop and rock performers.
SADC Calling	A program produced by the Southern African Development Community focusing on cultural, economic, political and social issues in the region.
SAfm Lifestyle	A masala mix of features highlighting the diversity of cultures and styles that make up South Africa.
SAfm Literature	Authors, editors, publishers, books, magazines, DVDs, internet publications along with a 30 minute weekly drama.
SAfm Sports Special	Coverage of South African and international sporting events.
Saturday AM	Australian national current affairs.
Saturday Extra	A lively array of stories and features covering a range of international topics, politics and business.
Saturday Live	A weekend feature magazine program.
Saturday Morning	A mix of current affairs, feature interviews, music and food.
Saturday Night	Music, reminiscences and entertainment, including listener music requests.
Saturday Night Blues	Holger Petersen with a broad spectrum of blues-based music, from Mississippi Delta blues to roots rock, zydeco and swing.
Saturday Night Country	The music, personalities and latest news from the Australian and international country music scene, hosted by John Nutting.
Saturday PM	The day's news and the big events in South Africa and around the world.
Saturday Review	Sharp, critical discussion of the week's cultural events.
Science in Action	A roundup of the latest news from the worlds of science and technology.
Science Plus	Estelle Winters with a wide range of themes related to scientific matters.
Shipping Forecast	The maritime weather forecast for the British Isles and vicinity.
Short Story	The best of classic and contemporary short fiction from Australian and world writers.
Something Understood	Examining the larger questions of live, taking a spiritual theme and exploring it through music, prose and poetry.
Songs from Russia	Musical novelties and melodies from the past.
Sound Czech	Learning useful Czech language phrases through song lyrics.
Sound Quality	Tim Ritchie with an eclectic mix of music.
Sounds Historical	Nostalgic news, features and interviews looking at New Zealanders the way we were.
Soundwaves	The latest in English and Chinese pop music.
Spark	A weekly audioblog of trendwatching, about the way technology affects our lives and the world around us, with Nora Young.
Speaking Out	A program about Australian Aboriginal and Torres Strait Islander people.
Special Delivery	A second shot of a popular CBC Radio series, documentary, feature or interview.
Spectrum (on DW)	A weekly magazine looking at developments in the fields of science and technology.

Program Name	Description
Spectrum (RNZ)	People, places and events in New Zealand
Spiritual Flowerbed	Russian clergy offer reflections on spiritual matters.
Sports Factor	A program that debates and celebrates the cultural significance of sport.
Sportsworld (on BBC)	All the international sporting action from around the world.
Spotlight	Touring the Czech Republic
Start the Week	An arts and culture program.
Stir It Up	Sampling the Swiss food scene with Anne Glusker.
Storytime	New Zealand stories for children.
Studio 15	Music and literature festivals around Poland.
Sunday	Religious news and current affairs.
Sunday Morning (CBC-R1)	Discussion, features and ideas for a New Zealand Sunday morning.
Sunday Morning (RTHK-3)	A blend of music and chat for Sunday brunch with Natalie Haughton.
Sunday Night Talk	A unique weekly program exploring the issues, events and people driving developments in religion, ethics, spirituality, popular culture,, values and beliefs in Australia.
Sunday Panorama	A review of top items from the past week's "News and Views" programs.
Sunday PM	An update of the day's news and big events in South Africa and around the world.
Sunday Profile	An in-depth analysis of the major news in Australia and around the world.
Sunday Worship	Diverse services of worship and religious expression.
Swiss By Design	Exploring Switzerland's contemporary culture, looking at art, architecture, products and people with Jennifer Davies
Tagata o te Moana	News from the Pacific.
Talking Jazz	The Polish jazz scene.
Talking Point	A topical commentary.
Talking Point	A closer look at the issues shaping the daily lives of people in the Czech Republic.
Tam Tam Express	A weekly examination of governance in Africa.
Tapestry	Mary Hynes with a thoughtful consideration of faith, spirituality and what it means to be human.
Te Ahi Kaa	Exploring the Maori world with its diversity of opinion, perspective, insight and experience.
Teen Time	Music, style and issues of interest and concern to teens with Alyson Hau
The Afternoon Reading	A short story or an abridged book, often by writers who are new to radio.
The Age of Persuasion	Terry O'Reilly examines marketing and how it permeates our lives, from media, art and language to politics, religion and fashion.
The Archers	A daily British drama serial since 1950.
The Ark	A program that seeks to shatter the usual perceptions about religious topics with Rachel Kohn.
The Arts	Reports on the rich cultural life of the Czech Republic.
The Arts on Sunday	Information and analysis from the world of books, arts and movies.
The Best of Jazz	"Mr. Smooth," Ike Phaahla, plays jazz.
The Biz	Business and finance in Poland.
The Book Reading	Readings of New Zealand novels, short stories and short dramas adapted for radio by Radio New Zealand.
The Book Show	Ramona Koval hosts a program for the discussion of everything related to the written word.
The Bottom Line	Insight into business from the people at the top.
The Breakfast Show	Humor plus music and a fresh take on what's happening around the Switzerland and the world.
The Choice	CBC listeners request their favorite interviews and documentaries from the network's archives.
The Christian Message from Moscow	The Russian Orthodox Church, Russian saints, church music and religious literature.
The Classifieds	Swiss phone-in to buy, sell and exchange merchandise.
The Current	Journalist Anna Maria Tremonti hosts a program offering insight, context and meaning to the unfolding events of the day.
The Daily Planet	Lucky Oceans with the world's diverse, traditional and innovative musics--jazz, blues, folkstyles, art music and more.
The Debaters	Part competition, part quiz show and part stand-up performance, comedians go toe-to-toe.
The Film Programme	Francine Stark interviews star guests on the latest cinema releases, DVDs and films on TV.
The Forum	Bridget Kendall presents a program that discusses and challenges big ideas.
The House	Kathleen Petty and CBC Radio take you behind the scenes of Canadian politics.
The Inner Voice	A weekly inspirational program exploring spiritual principles.
The Inside Track	Robin Brown with interviews and documentaries that highlight the issues and people of sport.
The Instant Guide	A weekly backgrounder to the people, places and ideas in the news.
The Kids	Youth culture and education in Poland.
The Late Story	Short stories in the small hours.
The Learning Curve	Libby Purves and guests discuss current education issues.
The Link	Aimed at connecting new immigrants to Canada and Canada to the world.
The Maple Leaf Mailbag	Ian Jones reads listener letters, answers questions and shares impressions of Canada from around the world.
The Margaret Throsby Interview	Conversation and music with a cultural theme.
The Material World	Quentin Cooper reports and interviews scientists on developments across the sciences.
The Media Show	Radio 4's weekly look at the media.
The Moral Maze	Michael Buerk chairs a lively debate examining the moral issues behind the week's news stories.
The Music Mix	Music, talk, live sessions and coverage of contemporary music and events.
The Music Show	A mix of music, interviews and information about the latest developments in music hosted by Australian composer Andrew Ford.
The National Interest	Peter Mares with the major issues of the week in Australia.
The Next Chapter	Shelagh Rogers talks with Canadian writers and authors.
The Night Air	An audio adventure in which ideas, sounds and music are remixed around a new theme each week.
The Night Time Review	An evening reprise of the best from that day's CBC programs "The Current" and "The Point".
The Other Heartlands	How history and landscape have shaped political allegiances in some of Britain's most remote constituencies.
The Philosopher's Zone	Exploring the big philosophical questions and arguments of the day.
The Point	Conversation, debate and chat about the events and issues of the day with Aamer Haleem.
The Radio 4 Appeal	Charitable causes in the United Kingdom.
The Rhythm Divine	Geoff Wood with music and song from the world's religious traditions, new religious practices and the fusion of religion and pop culture.
The Sampler	Nick Bollinger reviews the latest CD releases.
The Saturday Play	Thrillers, mysteries, love stories and detective fiction, as well as an occasional special series.
The Science Show	ABC (Australia) Radio's longest running program covering scientific issues, debates, events, personalities and discoveries.
The Spirit of Things	An adventurous examination of religion and spirituality with Rachel Kohn.
The State We're In	Human rights issues around the world.

Program Name	Description
The Strand	A daily program which takes you on a worldwide journey through the arts, culture and entertainment.
The Sunday Drama	The best of New Zealand's writing, acting and directing talent.
The Sunday Edition	Conversation and documentaries on everything from politics to pop culture in Canada and the world, hosted by Michael Enright.
The Vinyl Cafe	Music and stories from "the world's smallest record store," as conceived by the mind of Canadian humorist Stuart McLean.
The VOR Treasure Store	Programs on history or culture drawn from the Voice of Russia's archives.
The Wednesday Drama	The best of New Zealand's writing, acting and directing talent.
The Week on Three	Highlights from the past week's programs on RTHK-3.
The World at One	Britain's leading political program with a reputation for rigorous and original investigation.
The World in Sport	Highlights of the world's sporting week with emphasis on New Zealand and the Pacific.
The World Today (ABC)	Australia's comprehensive noontime news and current affairs program.
The World Today (BBC)	A roundup of world news plus regional stories.
Thinking Allowed	Laurie Taylor discusses the latest social science research.
This is Russia	Profiles of Russia's cities and regions, nature, culture, the arts, religion and the Russians themselves.
This Way Up	Exploring the things we use and consume.
Time to Travel	Karen Kay explores beautiful travel destinations all around the world.
Timelines	Estelle Winters with a relaxed look at life in Moscow.
Today	Since 1957, Britain's daily agenda-setting morning program.
Today in France	A short daily report on aspects of French life and culture.
Tonic	Tom Tamashiro embraces the extended family of jazz with a mix designed to prepare for the week ahead.
Total Rugby	The world of Rugby Union.
Touchstone	A series exploring diverse spiritual, moral and ethical issues and topics.
Tradewinds	Walter Zweifel compiles a weekly review of Pacific regional business and economic news and features.
UN Chronicle	Reports on the work of the United Nations and its agencies.
Verbatim	Charting the story of the 20th century through the voices of ordinary Australians.
Vinyl Tap	Music and stories from Canadian rock legend Randy Bachman covering the history of modern popular music.
Vogue	Style in China.
Voices	Imogen Lamb talks with notable personalities traveling to and through Paris.
Voices from Other Lands	Interviews and discussions on topics concerning China and its ties to the rest of the world.
Waiata	A showcase and celebration of Maori music and musicians.
Wayne's Music	Wayne Mowat presents a selection of tune too good to be forgotten.
Weekend AM Live	Breaking news, community and lifestyle features, current affairs.
What's Up Japan	Current topics in the news in Japan and around the world.
White Coat, Black Art	Dr. Brian Goldman with an original and provocative show demystifying the world of medicine.
Windows on the World	International public radio features and documentaries.
WireTap	Eavesdroping on Jonathan Goldstein's "phone calls" provide a unique, sometimes strange, but always amusing narrative.
With Great Pleasure	A guest presenter picks readings of their favorite poetry and prose.
Woman's Hour	Celebrating, informing and entertaining women with Jenni Murray and Jane Garvey.
Word of Mouth	John and guests tackle listener questions about anything to do with the English language.
World Football	The big issues from the world's biggest sport.
World Have Your Say	The daily interactive show where listeners set the agenda.
World in Progress	Anke Rasper with a look at development issues around the globe.
World Interactive	Using listener letters and questions to present various facets of Japanese society and people's lives.
World on the Move	A nature series highlighting the great animal migrations on this planet.
World Tracks	World music performers and performances from France.
World Update	The latest news from around the world.
World Vibes	Explore music from around the world with presenter Pierre Trembley.
Worldwatch	International news and reports.
Writers and Company	Eleanor Wachtel interviews writers from around the world.
You and Yours	Radio 4's consumer affairs program.
Your Space	Michele Mischler and Mark Butcher open the phone lines to discuss the current topics that impact the daily lives of the Swiss.

Making Sense of it All

If you are at all familiar with my last series of *Shortwave Listening Guides*, you'll know that there was somewhat of a challenge involved in putting a "fence" around what is the organized chaos of shortwave-based international broadcasting. The advent of new delivery platforms combined with the de facto expansion of the no-longer-easy-to-define-because-of-the-borderless-internet "international broadcaster" label, hasn't made the erection of that fence any easier. The proverbial "herding cats" reference applies here in spades.

The former bright line between international and national broadcasting has been blurred if not erased entirely by the broader reach of the internet relative to local and even international terrestrial broadcasting. While many services continue to characterize and market themselves as international or national, the fact that both can now be readily heard across borders allows listeners to determine how they each best work for their personal needs or preferences.

For the purposes of this book, it became quickly apparent that whether a service is expressly national or international is less important than whether or not the service is accessible internationally via some relatively convenient delivery platform. Obviously, a major non-technical factor in making a station or service accessible is language. For English speakers, the most valuable and accessible stations and services are those that broadcast entirely – or at least to a significant extent – in English.

Most international radio listeners also express a preference for public service broadcasting over commercial stations and services. The major international broadcasters, going back several decades to their beginnings, have been non-commercial, public or government supported in nature and character. We decided also to keep that focus in this new book.

In the listings section of this book, we initially set out by keying on the six primary English language countries, other than the USA, and their national public radio networks, regardless of delivery platform. For clarification, those nations are the United Kingdom, Australia, New Zealand, Ireland, Canada and South Africa.

United Kingdom

We include full listings for several of the **BBC World Service** (WS) program streams and **BBC Radio 4**. Admittedly, there is much more on offer – and in one way, much less.

The "less" has to do with shortwave. Simply put, the **BBC World Service** has consciously made itself very difficult to hear in North America on shortwave. It is still sometimes possible, but reception is far from reliable and, even then, only really possible with better-quality radios using an enhanced antenna system. Many see this eventuality as unfortunate. If better capable shortwave facilities are available to you, try these frequencies and time (UTC) frames:

Frequency	UTC
21470	1200-1700
17830	1100-2100
15400	1500-2300
12095	1500-1700
9740	0900-1600
7255	0300-0600
6005	0300-0700
5875	0500-0700

Since all of these frequencies target areas quite remote from North America, it bears repeating that reception is not assured. Also, the streams broadcast on these frequencies do not correlate to the listings in this book. For the most part, they will follow schedules designed for Africa and Southeast Asia. The WS web site www.bbcworldservice.com provides programming details.

BBC Radio 4 has a style and follows a schedule that most closely resembles the WS that most middle-aged and older listeners first found so compelling. That is why it appears in the listings along with schedules for the four WS streams available in North America via platforms other than shortwave.

If you're inclined to explore further with the internet, here are the other BBC national radio networks which are more thematic in nature:

- BBC Radio 1 – New music and youth network

- BBC 1Xtra – Urban music and culture

- BBC Radio 2 – Popular music and entertainment

- BBC Radio 3 – Classical, jazz, world music, arts, drama

- BBC 5 Live – Live sports and news

- BBC 6 Music – Alternative, indie, rock, dance music.

- BBC 7 – Archival material, mostly drama and comedies.

- BBC Asian Network – South Asian music and culture

Digging deeper, there are three popular "sub-national" radios:

- BBC Radio Scotland

- BBC Radio Wales

- BBC Radio Ulster

Drilling even further down, each of the country's major cities and regions have their own local radio stations, such as BBC Radio London, BBC Radio Manchester and BBC Three Counties Radio. To access the schedules and audio for all of the foregoing, start at **www.bbc.co.uk/radio** and continue in whatever direction you choose!

Here are some additional things to note about BBC Radio.

1. The BBC reserves its Windows Media streams for UK domestic audience use only. Real Media streams are available to both domestic and international audiences.

2. While BBC Five Live, and its overflow network, BBC

Five Live Extra, carry play-by-play sport commentaries, most of these are available only to its UK audience due to copyright restrictions and business considerations. There are exceptions, however. Many cricket, rugby and tennis matches are left "in the clear" for listeners outside the UK.

3. In recent years, the WS has made a concerted effort to secure carriage on a number of local stations around the world that also stream their content on the internet. For example, even though the small BBC Caribbean Service no longer broadcasts on shortwave, it is rebroadcast on a number of Caribbean stations that also stream on the internet. The WS provides a listing here:

http://www.bbc.co.uk/caribbean/institutional/frequencies.shtml

Use this information to perform a search on Google or Yahoo to link to the station web site and learn if it streams its content on the internet. This process can be repeated to find webcasts of other WS regional streams.

The UK also has a roster of successful national commercial radio stations. Among them are:
Classic FM (classical music) – **www.classicfm.com**
TalkSport (24 hour sports) – **www.talksport.net**
Virgin Radio (pop and rock music) – **www.virginradio.co.uk**
Gold (classic rock) – **www.golddigital.co.uk**

Australia

We include the schedules of two major ABC networks in the listings: the international service, **Radio Australia**, and the ideas and spoken word network, **Radio National**. The two networks share some programming. Only **Radio Australia** broadcasts on shortwave and does not intentionally target North America with its transmitters. Thus, the listings do not include any shortwave frequencies.

Nonetheless, these broadcasts do make their way to our shores, more reliably – but not exclusively – to the western half of the continent. Here are some suggested frequencies and times (UTC) that may periodically provide acceptable results:

Frequency	UTC
5995	1400-1800
6020	1100-1400
7240	1400-1700
9580	0800-1400
9590	0800-1600
9710	1600-2000
11880	1700-2100
13630	0700-0900
15160	0500-0800
15240	0000-0800
15515	0200-0700
17715	0000-0200
17785	2200-0000
17795	0000-0200

As with the BBC, the ABC has other thematic-based national networks that stream their content on the internet:

• ABC Classic FM – classical music

• ABC Triple J (JJJ) – rock music and youth culture

• ABC NewsRadio – continuous news, with supplemental content from the BBC World Service and World Radio Network (WRN)—Australia.

• dig – three continuous music networks (rock, jazz, country) exclusive to the internet.

ABC Local Radio operates as a national network with local programming variations. The stations in the network which stream audio on the internet are:
• ABC Coast FM
• ABC Brisbane
• ABC Newscastle
• ABC Sydney
• ABC Melbourne
• ABC Hobart
• ABC Adelaide
• ABC Perth
• ABC Darwin
• ABC Canberra

ABC Local Radio's weekend "Grandstand" program provides extensive and live coverage of the country's sporting events, but as with the BBC, much of this content including the widely popular local favorites rugby and Australian rules football cannot be streamed on the internet to an overseas audience due to rights restrictions. These restrictions do not apply to the shortwave broadcasts! However, cricket is featured extensively both on the radio and over the internet (**ABC Cricket**) and special coverage is made widely available for iconic national events such as the Australian Open tennis championships and the Melbourne Cup Spring Racing Carnival. To explore schedules and links, refer to **www.abc.net.au/radio**.

Coverage of Australian sport also is available over the internet from the Melbourne commercial stations **Sport 927 www.sport927.com.au/sport927/**, which features extensive horse and greyhound racing commentary and **SEN 1116 www.sen.com.au/** which has a much wider focus.

New Zealand

This book includes full listings for **Radio New Zealand International** (RNZI) and **Radio New Zealand National**. When RNZI is not broadcasting its own content, it relays the National network. RNZI shortwave broadcasts are not intentionally targeted to North America. Nonetheless, their frequencies are regularly heard quite well here (again, better on the western half of the continent) and are included within the listings.

Also streamed over the internet is **Radio New Zealand Concert**, a classical music network. Refer to **www.radionz.co.nz/concert**

Sports are as rabidly popular in New Zealand as it is in Australia. In years past, RNZI carried weekend play-by-play rugby commentary from the commercial network **Radio Sport www.radiosport.co.nz/** . It no longer does so, but a wealth of national and international rugby and other sports coverage is available to listeners from Radio Sport internet streams. Some overflow appears on its sister commercial network **Newstalk ZB www.newstalkzb.co.nz/** which also serves, as it name indicates, as a major national news and talk network.

Radio Hauraki www.hauraki.co.nz/ , a commercial pop and rock radio station, has a colorful history dating back to its birth as a pirate station housed on a ship anchored offshore, recalling similar "outlaw" operations around the UK, before commercial radio was a government-sanctioned activity.

Ireland

This book includes full listings for the eclectically programmed **RTE Radio 1**. In addition, RTE Radio **www.rte.ie/radio** programs several networks that hew to more disciplined themes. They are:

- 2fm – Rock and pop music.
- lyric-fm – Classical music and jazz.
- Raidio na Gaeltachta – Gaelic language network.
- 2XM – Continuous rock music.
- Gold – Music of the '50s, '60s and '70s
- Junior – Children's network
- Digital Radio News – Continuous news.
- Pulse – Dance music.

In addition, there are two national commercial networks: **Today FM www.todayfm.com** , which programs primarily popular music and **Newstalk www.newstalk.ie** , which is precisely as its name suggests.

Canada

Once known as the CBC International Service, **Radio Canada International** has evolved from a service with a wide remit to explain Canada to foreign listeners, to one which, for a time, simply relayed domestic radio content to its English speaking listeners. Today, RCI has a much more focused primary audience target – immigrants and potential immigrants to Canada. This singular mission is underlined by the fact that RCI no longer relays any CBC domestic programs. Fortunately, RCI's programming still works for a wider audience and its full shortwave (to North America) and internet radio schedules are included in this book's listings.

By virtue of its five (and a half! – Newfoundland) time zones, **CBC Radio 1** provides an expansive schedule of wide-ranging programs. CBC and SiriusXM satellite radio have a working relationship whereby a separate **CBC Radio 1 on Sirius** schedule is provided. All six of these schedules are included in this book.

In addition, CBC Radio **www.cbc.ca/radio** offers these more specifically focused services:

- CBC Radio 2 – Classical, jazz, Canadian arts and performances.
- CBC Radio 3 – Canadian new music and artists.
- Four internet only music genre channels – Classical, Jazz, Composers, Songwriters.

South Africa

This book provides the full published schedules of the South African Broadcasting Corporation's international service, **Channel Africa www.channelafrica.org** , and its main domestic English language network, **SAfm www.safm.co.za** . Both are streamed over the internet. Channel Africa also broadcasts on shortwave, but only to the African continent. Therefore, its shortwave frequency schedule is not included within our listings.

Where To From There?

From the starting point of these six nations and their broadcasting services, we sought to gradually widen the focus to longstanding, strong international services from other countries as well as stations in pockets of nations with enough of a significant English language presence to merit services in that language.

As an example, Hong Kong as a former British colony, clearly meets this criteria and we've included **RTHK Radio 3** (Radio-Television Hong Kong) in this book's listings inasmuch as it is an eclectic, full-service radio station for the English speaking community there and offers a further glimpse of life in at least one part if not more of China. Pairing the internet availability of **RTHK Radio 3** with the very strong shortwave and 24 hour a day internet presence of **China Radio International** (CRI) can provide the attentive and interested listener with a degree of insight to life there that might not be available through any other media.

We came across some surprises in pursuing this line of inquiry. **World Radio Switzerland** is a 24 hour a day service for English speaking residents of Switzerland, as part of the Swiss Broadcasting Corporation. While nowhere near as satisfying as the national broadcaster's former Swiss Radio International, it nonetheless fills some of that void with news and views from the Swiss cantons, albeit the focus is internal rather than SRI's external remit. (Those who miss SRI as much as I do might wish to frequently sample the web site **www.switzerlandinsound.com** constructed and maintained by Bob Zannotti, one of the famous "Two Bobs" from that popular service.)

Another European surprise is **ORF-fm4** in Austria. This service is bilingual – German and English – with a part of each broadcast day given over to predominant use of English. Program schedules for both WRS and ORF-fm4 are included in the listings section of this book.

News from Italy in English (but unfortunately not any more than that) can still be obtained through the internet via the **RAI International** overnight (in Europe) service, *Notturno Italiano*. Slovenia cannot be heard at all via shortwave, but its national broadcaster has created a multilingual international service titled **Radio Slovenia**

International **www.rtvslo.si/rsi/** that streams over the internet, as well as broadcasts locally in Slovenia on FM and AM for English speaking visitors and expatriates working in the country. Its programming consists mostly of pop and rock music programs. For that reason, RSI is not included in this book's listings. But we thought it merited mention here. A similar service is Malaysia's **Traxx FM www.traxxfm. net/** , which is the country's sole full time English language network but is largely a pop music station. However, you may wish to tune into its internet stream for a newscast or talkback show to gain a perspective on or from Malaysia.

Some longstanding international services are still straddling the ground between shortwave and the internet. In addition to the aforementioned CRI, the **Voice of Russia**, and **Radio Prague** fall into this category. We have included both their shortwave broadcast (to North America) and internet streaming schedules in the book's main listings. In each case, the internet schedule represents a significant value-added opportunity for their listeners.

A Strong Migration to the Internet

Longtime devotees of shortwave broadcasting will note that the use of it as a delivery platform to North America is markedly less prevalent than it was even a few short years ago. That pattern also is represented in the approaches of several of the prominent English language international services.

Deutsche Welle and **Radio Netherlands** were dominant broadcasters on shortwave to North America for decades. Neither targets its shortwave broadcasts to this continent any longer. Both, however, are streaming content twenty-four hour a day over the internet and their full schedules are provided in this book. **Radio France International** is another popular international broadcaster that some time ago left North America behind when it comes to shortwave delivery. Its English Service represents this important nation well and maintains a strong presence on the internet. Its published schedule is in the listings section.

On the other hand, **Polish Radio External Service** has not ever (or at least in recent memory) targeted North America with shortwave transmitters. However, its well-produced, pleasant and informative programs provide valuable insight into and perspectives from eastern Europe. The entire internet streaming schedule also is included herein.

Where To From Here?

Although great pains were taken in constructing this new edition to maintain a logical thread throughout, attentive readers will still note some gaps. At several points, the sheer volume of potential material threatened to overwhelm the effort. So, on one level, a conscious decision was made to attempt to identify and include the most popular and demonstrably prominent stations and networks. In other cases, no published or detailed programming information was readily available and efforts to secure it by other means did not meet with success in time for inclusion in this edition. In the case of U.S broadcasters, we reasoned that most readers/listeners of this book would be seeking an international perspective. So, for this edition, reference to North American AM and FM stations are confined to those that provide themselves – or relay other stations that provide – those perspectives. Finally, in still other cases, stations were deemed well enough known and with less diversified schedules that were easily obtainable by the listener from other sources or by other means; and, therefore, inclusion here would be of lesser benefit than inclusion of others.

Admittedly, this was a somewhat subjective process and observers are free to disagree with the judgments made by the author. For his part, the author freely acknowledges falling short in some respects, even of his own hopes and expectations. Readers should feel free – indeed, you are most cordially invited – to share impressions, constructive criticism and suggestions for future editions. This should be seen as an organic process that will improve incrementally as these new delivery platforms and broadcasting systems develop and become better understood.

Nonetheless, even with acknowledged shortcomings, the author is confident that listeners will find this resource useful, unique and sufficiently flexible in structure to permit the reader to supplement and correct information in a manner that better tailors it to individual preference and need.

Listener's Log

Date	Time	Station	Listener's Notes

Listener's Log

Date	Time	Station	Listener's Notes

Listener's Log

Date	Time	Station	Listener's Notes

Listener's Log

Date	Time	Station	Listener's Notes

Listener's Log

Date	Time	Station	Listener's Notes

Listener's Log

Date	Time	Station	Listener's Notes

Listener's Log

Date	Time	Station	Listener's Notes

Listener's Log

Date	Time	Station	Listener's Notes

Listener's Log

Date	Time	Station	Listener's Notes

Listener's Log

Date	Time	Station	Listener's Notes

Listener's Log

Date	Time	Station	Listener's Notes

Listener's Log

Date	Time	Station	Listener's Notes

Listener's Log

Date	Time	Station	Listener's Notes

Listener's Log

Date	Time	Station	Listener's Notes

Listener's Log

Date	Time	Station	Listener's Notes

Listener's Log

Date	Time	Station	Listener's Notes

Listener's Log

Date	Time	Station	Listener's Notes

Listener's Log

Date	Time	Station	Listener's Notes

Listener's Log

Date	Time	Station	Listener's Notes